Second Edition

Biological Inquiry
A Workbook of Investigative Cases for

Campbell • Reece

Biology

Eighth Edition

Margaret Waterman
Southeast Missouri State University

Ethel Stanley
BioQUEST Curriculum Consortium and Beloit College

San Francisco Boston New York
Cape Town Hong Kong London Madrid Mexico City
Montreal Munich Paris Singapore Sydney Tokyo Toronto

Editor-in-Chief: Beth Wilbur
Senior Editorial Manager: Ginnie Simione Jutson
Senior Supplements Project Editor: Susan Berge
Project Editor: Kim Wimpsett
Managing Editor: Michael Early
Production Supervisor: Jane Brundage
Photo Researcher: Maureen Spuhler
Photo Editor: Donna Kalal
Manufacturing Buyer: Michael Early, Michael Penne
Executive Marketing Manager: Lauren Harp
Production Management and Composition: S4Carlisle, Judy Ludowitz
Cover Designer: Yvo Riezebos Design
Text and Cover Printer: Technical Communications Services

Cover Image: Magnolia Flower–Corbis. Photographer: Chris Fox.

ISBN-13: 978-0-321-51320-5
ISBN-10: 0-321-51320-7

1 2 3 4 5 6 7 8 9 10 – TCS – 10 09 08 07
www.aw-bc.com

Dedication

We dedicate this book to the memory of Keith D. Stanley (1949–2004), an extraordinary starch chemist, and to the memory of Neil Campbell (1946–2004), an extraordinary biology teacher and author. These two gentle men never lost their sense of wonder in this world.

Preface

Biological Inquiry: A Workbook of Investigative Cases is a book of short cases (stories or scenarios) based on decisions people face in everyday life that require familiarity with biological concepts. Each case is designed to complement a related unit in *Biology*, 8th edition, by Campbell and Reece. These investigative cases provide opportunities for you to apply the biology you are learning in the classroom to realistic situations.

Your instructor may use the cases in a variety of ways during your course. For example, the cases could be assigned as homework, integrated into a lecture, or incorporated into a laboratory. You might work alone or in groups. Working in groups can give you insight into the way scientists work because real-world scientific investigations often involve collaboration. Even if you are not planning on becoming a scientist, it is important to understand the way scientists work in order to better grasp biological issues that affect you as a global citizen.

Regardless of the way your instructor chooses to use this book, this preface will give you some general advice and insight into learning with cases.

(1) What is a case?

A case is a type of scenario that is useful for learning. In general, cases are created in many formats including videos, computer-based programs, and written forms. Text-based cases, such as the ones in this book, are common and they can be one paragraph or many pages long. In this book, the cases are about a page long.

Following is an example of a short biology case similar to ones you will find in this book.

Case: *Derrick's Malaise*

About a month after returning home from a season of fieldwork in Guatemala, Derrick began to feel sick again. His roommate took him to the clinic. They were both worried that he was having a recurrence of the malaria he had contracted on the trip.

"How could I have these symptoms again?" Derrick asked the resident, Dr. Welty. "I finished the prescription they gave me in Guatemala and I have been feeling fine."

"I'm not sure why you have this again," Dr. Welty replied. "We'll need a blood sample so we can analyze the organism to see what strain it is. In the meantime, let's try a different drug. I'm going to switch you to chloroquine." He continued, "You know, malaria is one of the most common infectious diseases in the world. I've got some contacts at the CDC who may be interested in your relapse. May I share your records with them?"

"I wonder if I can get malaria from Derrick," wondered Derrick's roommate, who was flipping through magazines in the waiting room.

Case Authors: Ethel Stanley and Margaret Waterman, 2001. Investigative Cases and Case Based Learning in Biology, version 1.0, in Jungck and Vaughan (Eds.), BioQUEST Library VI. San Diego: Academic Press.

(2) How do I begin?

Begin by finding out what the case is about. Read through the case to get a sense of the story and issues. If you are working in a group, try having one person read the case aloud while the others read along silently. This may sound unusual, but it helps everyone in the group focus on the case.

(3) What is the Case Analysis all about?

Once your group has read the case, go to the Case Analysis sheet found just after each case. Case Analysis helps you to identify the main ideas in the case, as well as what you already know about the situation, and what your questions are about the case. If you analyze the case in a group, you will share your ideas, hear what others are thinking, and have a good sense of what the group needs to learn about this case. Case Analysis involves four steps.

Step A. Recognize potential issues and major topics in the case.

Go back and read the case again, this time highlighting or underlining words or phrases that seem to be important to understanding the situation. Look for issues that you might explore further. Jot down your ideas and questions about these words and phrases. If you are working in a group, this approach might be done as a group discussion, with one person keeping a list of issues as they are also raised. At this point, you are also answering the question, "What is this case about?"

The following is an example of the kinds of issues raised in the case *Derrick's Malaise.*

What does the case seem to be about? Malaria, how it is caused, why it might reoccur, how it is treated, and how it is transmitted.

What are some potential issues? How malarial drugs work, Derrick's relapse, worldwide rates of malarial infections, the role of the CDC.

Step B. What specific questions do you have about these topics?

In this step of Case Analysis, you will share what you already know, or what you think you know, and you will raise your questions. The "Know/Need to Know" chart, found in each Case Analysis, is a way to organize your thoughts. An example is included below.

Use Case Analysis as a brainstorming session. You can refer back to the underlined words and phrases in the case as a way to help organize this discussion. This step can be accomplished alone; however, experience shows it is better done in a group.

Using the *Derrick's Malaise* case as an example, here are some questions raised by learners who have worked with this case:

What Do I Know?	What Do I Need to Know?
• Malaria is transmitted by mosquitoes. • It is found in locations that are warm and damp. • It isn't common today in the United States. • It can be treated with drugs. • You can get better. • It can recur (from the case). • It is caused by a microorganism. • The CDC is the Centers for Disease Control and Prevention. • Many people die of malaria each year.	• Why was Derrick in Guatemala? • How common is malaria—worldwide, Guatemala, United States? • What role does the CDC have? • What kind of organism causes malaria? • Should Derrick's roommate be concerned? • What is chloroquine? Is it a common drug? How does it work? Is it safe? • What other drugs are used? • How do you prevent malaria? • Do all types of mosquitoes transmit malaria? • What strains of the malarial organism are there? • Is it okay for the doctor to switch drugs without knowing more?

Step C. Assign priority to the questions.

Review the questions listed on the "What Do I Need to Know?" side of the chart. It is very likely that your brainstorming session raised many different kinds of questions on many topics related to the case (but not necessarily to biology). Go over your list and put a check by the three questions that seem most important to understanding this case. One way to do this is to think about which questions fit with the topics on your course syllabus or in the textbook chapters being studied. Check off those questions as well as others that interest you the most.

> If *Derrick's Malaise* was introduced while you were reading a chapter on microorganisms, questions about the organism causing malaria would be fairly important to investigate. Other questions that you find interesting but that are not linked to a syllabus topic may also be chosen for study, such as "How does chloroquine work?"

You will find that some of your questions may be addressed in the investigations that accompany the case. You might also have questions that this Case Book does not address. Your instructor might suggest that you expand on these questions by developing a paper, presentation, experiment, ethics statement, or other product.

"One of the greatest challenges in biology is to frame appropriate and productive questions that can be pursued by the technology at hand. You have probably had a great deal of experience in solving pre-posed problems, such as those found at the end of textbook chapters. However, if you were asked to go into a lab or out in a field and pose a research question, you will find that this is often difficult to do without some practice . . ."

(The BioQUEST Library IV: *A Note to the Student,* University of Maryland Press, 1996)

Step D. What kinds of references or resources would help you answer or explore these questions?

No matter what questions you investigate, it is likely you will seek and use resources to help you develop persuasive answers. It is important to develop the habit of thinking broadly about where you might go to find answers to your questions. Resources may include your textbooks, other library materials, computer simulations, results of lab or field research, articles, data sets, maps, e-mails, pamphlets from organizations, interviews with experts, or museum exhibits. Be creative, but remember your data are only as good as your sources.

For *Derrick's Malaise,* you might choose to examine:

- Maps of malaria prevalence
- A reference book such as *Physicians' Desk Reference* to find out how chloroquine works
- A Web page from the CDC with international travel precautions
- Your textbook index for relevant terms
- A simulation in which the prevalence of different species of competing mosquitoes is examined under different conditions
- An interview with a person who has had malaria

You will find links to a variety of online resources referenced in the cases on the Campbell website (http://www.campbellbiology.com) under the tab marked "Case Book." In addition to the resources organized by case, you will find additional open-ended investigations where you can pose your own questions.

(4) What do I do after the Case Analysis?

Each case is accompanied by several investigations. Your instructor might assign just one or two of the available investigations, depending on what fits best in your course. You can also use this book independently. You might complete the Case Analysis and selected investigations as a way to check your own knowledge.

(5) What if my instructor wants me to develop my own investigations?

Your group is likely to raise different questions in the case analysis from the ones investigated in this book. Sometimes an instructor might ask you to follow up your own question in a lab or to design an investigation of your choice. Following is some advice on ways to turn your case-related questions into scientific investigations.

A. Getting started: How do I develop my question?

As you develop the problem and questions you want to use to investigate and learn more about the topics, it will be important for you to consult with others, such as members of your group or other classmates. Discussing your ideas and plans is an important step in refining problems and can lead you to different perspectives and possible good research problems. Continue this practice of sharing with others as you gather evidence for your problem and as you prepare to present your conclusions. This kind of communication is the standard among scientists.

B. What am I expected to do with my question(s)?

Once you have a problem you want to investigate, you and your instructor might consider any of the following:

- Design and conduct new investigations utilizing laboratory or field methods.
- Use computer software modules, spreadsheets, simulations, data sets, interactive maps, remote sensing, or graphics to investigate the question.
- Seek new sources of data (further references, interviews, data sets).
- Develop an investigation that builds from a standard lab exercise, perhaps by changing the independent variable or establishing new controls.

Following are three possible investigations for *Derrick's Malaise.*

1. Work with a simulation to investigate hypotheses about control of mosquitoes that carry malaria-causing organisms. Nonvector mosquitoes can be introduced to compete with the vectors.
2. Use genomics tools and the PlasmoDB (a database of gene sequences from many species of *Plasmodium*) to examine genes in resistant and nonresistant strains of *Plasmodium falciparum.*
3. Develop an experiment to test the effectiveness of controlling mosquito populations with fish that feed on larvae.

C. When am I finished?

> "... You must confront the issue of closure in research. How do you know when you have a 'right' answer? When is research done? Scientists do not arrive at a final answer; usually research is abandoned for a variety of reasons, including time, resources, and most importantly, when the scientific research team is 'satisfied' with their conclusions, that is, when the solution is 'useful' for some purpose."
>
> (The BioQUEST Library IV: *A Note to the Student,* University of Maryland Press, 1996)

When you are ready to present your conclusions, remember that you need to persuade others of the value of your methodologies and data. Consider your audience carefully as you develop products to support your conclusions, such as

- Scientific posters
- Advertisements urging political action
- Videos defining the issues for the public
- Pamphlets/brochures with recommendations for a specific user group
- Consulting reports (if you are role-playing)
- Art work, such as cartoons, revealing issues from the case
- Designs for a new technological approach to the problem
- Scientific reports to local or regional groups
- A new case study to emphasize your findings

> "Research is not complete, no matter how many experiments have been conducted, no matter how many puzzles have been solved, until peers outside of a research team are persuaded of the utility of the answers. Persuasion is a social process and an essential one for you to experience in order to understand the nature of scientific theories and paradigm shifts. Communication in the science community is an active process full of controversy and debate. The productive side of science involves open criticism of the methods and conclusions made by a research group. This controversy and debate is important to the creation and acceptance of new scientific knowledge."
>
> (The BioQUEST Library IV: *A Note to the Student,* University of Maryland, 1996)

D. How will our group work be assessed and evaluated?

Like many students, you probably have concerns about the assessment and evaluation methods used in group work, especially in scientific inquiry.

Peer review is a key feature of how scientists judge each other's work. With investigative cases, you are likely to peer review one another's proposals, investigations, and persuasive materials. Recently, self-assessment has become a more frequent component of assessment in science, especially as more group work is done.

There are many ways to assess group products and group processes. Some instructors give a group grade and an individual grade. Other instructors include either peer evaluations or group self-evaluations in the grading process. If your instructor has not already explained how you will be assessed, you might want to discuss this.

(6) Why are cases a good way to learn biology?

An important goal of biology education is that you be able to apply what you learn in courses to your life. Scientific problem solving is a valuable tool in both professional and everyday life. It is important for you to do science as well as learn about it; and it is important for you to choose the problems to be studied and the resources you will use as you investigate the case.

By doing investigative cases, you engage in scientific inquiry. You will read critically, pose questions, analyze data, think critically, construct hypotheses, investigate options, interpret results, and communicate scientific arguments. No matter how your instructor chooses to use this book, investigative cases can be a useful and interesting tool in your study of biology.

Acknowledgments

In our second edition, we wish to reaffirm that books are the products of many people. Our extended community of support should now also include Ginnie Simione Jutson, senior editorial manager; Susan Berge, senior supplements project editor; Kim Wimpsett, project editor; Michael Early, managing editior; Michael Early and Michael Penne, manufacturing buyers; Jane Brundage, production supervisor; Maureen Spuhler, photo researcher; Donna Kalal, photo editor; Lauren Harp, executive marketing manager; and Judy Ludowitz and S4Carlisle, production management and composition.

A book is the product of the efforts of many people, not just the authors. For developing the concept of the book with us and supporting us without hesitation, we thank our friend and colleague Robin Heyden. Neil Campbell helped us shape the book originally, and both he and Jane Reece gave us unwavering support.

The editorial team at Benjamin/Cummings is without equal. We thank Beth Wilbur (editor-in-chief) as the leader of the team. Her willingness to hear our ideas and to let us go in new directions has energized our creative work. In the previous edition the day-to-day work on the book was managed professionally and congenially by Amy Austin, our project editor. Her commitment to clarity, her management of the many components of the book, and her thoughtful integration of reviews had a tremendous impact on the final version. There are many others at Benjamin/Cummings who had a role, including Vivian McDougal, production supervisor; Jeff Hester, biology marketing manager; and Kristin Piljay, photo researcher.

We thank the many academic biologists who reviewed the chapters in this book. Not only did they teach us a lot of biology, but they were tremendously positive about developing a case book. They include some people who are longtime collaborators, such as Peter Woodruff at Champlain College, Linda Weinland at Edison College, and Sam Donovan at the University of Pittsburgh. They also include people we have just met through this project who willingly gave their time and thoughtful reactions: Nan Arens, Hobart and William Smith Colleges; Jeanne Barnett, University of Southern Indiana; Ruth Buskirk, University of Texas–Austin; Joyce Cadwallader, Saint Mary-of-the-Woods College; Eloise Carter, Emory University; Christine Case, Skyline College; Dennis Clegg, University of California–Santa Barbara; Maura Flannery, St. John's University; Allan Gathman, Southeast Missouri State University; Dan Gleason, University of Oregon; Jeff Hardin, University of Wisconsin–Madison; Jean Heitz, University of Wisconsin–Madison; Tim Judd, Southeast Missouri State University; Graham Kent, Smith University; Stacey Kiser, Lane Community College; Janet Lanza, University of Arkansas–Little Rock; Lisa Meffert, Rice University; Michelle Mischke, Massachusetts Institute of Technology; Manuel Molles, University of New Mexico; Gary Munkvold, Pioneer Hi-Bred; Kevin Padian, University of California–Berkeley; Laura Olsen, University of Michigan; Peter Russell, Reed College; Stuart Sumida, California State University–San Bernardino; Lisa Urry, Mills College; Brian White, University of Massachusetts–Boston; and Brad Williamson, Olathe School District, Kansas.

In the second edition, we were also helped by the review, commentary, and advice given by the following colleagues: Amanda Everse, BioQUEST; Stephen Everse, University of Vermont; Marian Fass, Beloit College; Christina Frazier, Southeast Missouri State University; Amee Godwin, Institute for the Study of Knowledge Management in Education (ISKME); Claire Hemingway, Botanical Society of America; John Jungck, Beloit College; David Matlack, Earlham College; Patricia Marsteller, Emory University; Claudia Neuhauser, University of Minnesota; Felipe-Andres Ramirez-Weber, San Francisco State University; Anton Weisstein, Truman State University; and our students.

We are grateful to our colleagues at the BioQUEST Curriculum Consortium at Beloit College and Southeast Missouri State University who lightened our usual loads to give us the gift of time as well as funding for the required travel. We developed the idea for Investigative Case–Based Learning with the support of the National Science Foundation (NSF DUE 9952525), as well as with funding to BioQUEST from the Howard Hughes Medical Institute and the Education and Outreach Training Partnership for Advanced Computational Infrastructure. Although the funding support was theirs, any opinions, findings, conclusions, or recommendations expressed in this book are ours and do not necessarily reflect the views of NSF, HHMI, or EOT-PACI.

Our husbands, Keith Stanley and Steven Hoffman, fully supported our efforts. Keith Stanley, a carbohydrate chemist, collaborated on both *Picture Perfect* and *Galloper's Gut.* He graciously allowed us to take over the home office for the long term. We miss his quiet presence. Steven Hoffman, a historian, gave us advice on the museum elements of *Picture Perfect,* and willingly took over in Cape Girardeau while we were traveling and writing. Both of us are fortunate to have chosen to marry strong men who take pride in our accomplishments and see themselves as our partners. Russell Hoffman remained positive, helpful, and proud of what we are doing—even though his mom was away so much. Our children and grandchildren were in our thoughts every day (and sometimes in the book). Finally, we wish to thank our "extended" family who supported our work on this book in so many ways.

Ethel Stanley, Beloit, Wisconsin
Margaret Waterman, Cape Girardeau, Missouri

Credits

Fig. 1.5: Figure adapted from www.ftns.wau.nl/agridata/starchpackfoam.htm.

Figs. 2.2 and 2.3: From E. D. Stanley et al., "Modeling wine fermentation," pp. 85–92 and software on CD-ROM. In J. R. Jungck et al., *Microbes Count!* Beloit, WI: BioQUEST Curriculum Consortium and American Society for Microbiology Press, 2003. Used with permission.

Figs. 4.2 and 4.4: W. Herb Wagner. "Dendrogrammaceae." In E. D. Stanley, *Visual Data Sets,* BioQUEST Library VI, San Diego: Academic Press, 2001. Used with permission.

Fig. 4.6: From H. A. Ross et al., DNA surveillance: Web-based molecular identification of whales, dolphins, and porpoises. *Journal of Heredity* 94:111–114, 2003. Copyright © 2003 American Genetic Association. Used with permission.

Figs. 4.8 and 4.10: David Hornack and Sam Donovan, from http://bioquest.org/bedrock/problem_spaces/whippo/background.php. Copyright © 2004 BioQUEST Curriculum Consortium. All rights reserved. Used with permission.

Fig. 5.1: www.scotese.com/gzelclim.htm. © 2000, PALEOMAP Project.

Fig. 6.5: From D. J. Caffrey, *The European Corn Borer Farmers' Bulletin,* FB1046, USDA.

Fig. 6.8: Adapted from art provided courtesy of National Corn Growers Association.

Table 6.1: From http://www.uky.edu/Agriculture/Entomology/entfacts/fldcrops/ef118.htm. Courtesy of Ric Bessin, Extension Entomologist, University of Kentucky.

Fig. 7.3: Adapted from M. E. Ensminger and C. G. Olentine. *Feeds and Nutrition.* Clovis, Calif.: The Ensminger Publishing Company, © 1978.

Fig. 8.4: From http://mddnr.chesapeakebay.net/eyesonthebay/index.cfm. Map used with the courtesy of the Maryland Department of Natural Resources. www.dnr.maryland.gov.

Fig. 8.5: From http://www.cbrsp_toc_mb_chl_page.htm. This site is hosted by the Horn Point Laboratory at the University of Maryland Center for Environmental Science. Used with permission.

Figs. 9.2a: Data Source: World Organisation for Animal Health (OIE) and national governments. Map production: Public Health Mapping and GIS Communicable Diseases (CDS). Adapted from World Health Organization. © WHO 2007. All rights reserved.

Figs. 9.2b: Data Source: WHO/Map Production: Public Health Mapping and GIS Communicable Diseases (CDS). Adapted from World Health Organization. © WHO 2007. All rights reserved.

Fig. 9.2c: Adapted from *California Agriculture*, University of California, July–September, © 2006. The Regents of the University of California.

Figs. 10.2 and 10.3: Adapted from M. Athar et al. Inhibition of smoothened signaling prevents ultraviolet B-induced basal cell carcinomas through regulation of Fas expression and apoptosis. *Cancer Research* 64:7545-7552, figs. 2A, 2C, 2004. http://cancerres.aacrjournals.org. Copyright 2004 by the American Association for Cancer Research.

Fig. 10.4: From J. B. Weitzman. Research news: Agonizing hedgehog. *Journal of Biology* 1:7, fig. 1, 2002. jbiol.com/content/1/2/7. See terms of BioMed Central open access license agreement at biomedcentral.com.

Fig. 10.5: From Stem cell basics: What are the potential uses of human stem cells and the obstacles that must be overcome before these potential uses will be realized? In *Stem Cell Information*, Bethesda, MD: National Institutes of Health, U.S. Department of Health and Human Services, 2006. http://stemcells.nih.gov/info/basics/basics6.

Photo Credits

Chapter 1 opener and Fig. 1.1: Andy Crawford © Dorling Kindersley, Courtesy of the History Museum, Moscow. **Fig. 1.4a:** Andrew Syred/Science Photo Library. **Fig. 1.4b:** Dr. Gary Gaugler/Photo Researchers, Inc. **Fig. 2.1:** Tom Wagner/CORBIS SABA. **Fig. 3.1:** Robert Longuehaye, NIBSC/Science Photo Library. **Figs. 4.1 and 4.5:** Flip Nicklin/Minden Pictures. **Fig. 5.2:** Colin Keates/Dorling Kindersley. **Figs. 5.3, 5.4, and 5.5:** Rick Schrantz. **Fig. 5.UN:** Ludek Pesek/Science Photo Library. **Fig. 6.1:** Daniel Esgro Photography/Getty Images. **Fig. 6.2:** Visuals Unlimited. **Fig. 6.3:** Andrew Syred/Science Photo Library. **Figs. 6.4a, b:** Dr. R. L. Nielsen/Purdue University. **Fig. 7.1:** EyeWire Collection/Getty Images Photodisc. **Figs. 7.5a, b, and c:** Nancy Atsumi. **Fig. 7.7 (left):** Philip Dowell/Dorling Kindersley. **Fig. 7.7 (right):** Colin Keates/Dorling Kindersley, Courtesy of the Natural History Museum, London. **Fig. 8.1:** Paul Johnson/Index Stock Imagery. **Fig. 9.1:** Stanley family records, privately held, permission granted. **Fig. 9.5:** The National Archives. **Fig. 10.1:** Bluemoon Images/Photolibrary.

Contents

CHAPTER 1

Picture Perfect

Picture Perfect

As she drove back to the museum, Bryn considered the box and the tiny dress it contained. It had been worn by a child in a 19th-century portrait of a local family already owned by the museum. Discovered in a trunk in an unheated barn by descendants, the dress was in surprisingly good condition.

Once she arrived, Bryn went to the curators' workroom to give the dress to Rob, the museum's textile conservator. Seeing Rob working intently, she quietly knocked on the half-open door. He put down his tools and looked up.

"Rob," she said excitedly. "Here is the dress I told you about from the painting! The donor was about to have it cleaned, but I'm so glad he called here first."

"You're not kidding! It's easy to ruin old fabrics," Rob said as he accepted the box with the tissue-wrapped dress. After putting on gloves, he unwrapped the old dress carefully and laid it flat on a clean table to examine it. He saw that the cotton dress was slightly yellowed and there was a small, stiff stain near the neckline. He wondered if that spot might prove troublesome. "This is terrific, Bryn. I'll do my magic, and with luck these discolorations and spots should disappear."

Bryn laughed. She knew that Rob's work had nothing to do with magic or luck. As she left the workroom, Rob grabbed an *Object Description and Restoration* form and began to fill it out in pencil. Next he gently brushed the dress. Using a metal probe, he scraped the stain at the neckline and placed the sample on a microscope slide.

Rob examined the slide with the microscope, noticing several granules mixed in with a few longer fibers. He was not surprised to see long cellulose fibers, which he knew to be cotton. The granules, though, which were smooth and oval-shaped with a diameter of about 75 μm (micrometers), came from the stain itself. He added a drop of a weak, yellowish iodine solution to the slide. The granules turned dark blue. Under *Treatment Plan* he wrote "Neckline stain: use amylase cleaning solution"—an enzymatic solution specific for removing starch.

Figure 1.1 Museum textile conservators apply their knowledge of the structure and function of macromolecules to clean, restore, and preserve old fabrics such as this dress.

CASE ANALYSIS

1. **Recognize potential issues and major topics in the case.** What is this case about? Underline terms or phrases that seem to be important to understanding this case. Then list **three or four** biology-related topics or issues in the case.

2. **What specific questions do you have about these topics?** By yourself, or better yet, in a group, list what you already know about this case in the "What Do I Know?" column. List questions you would like to learn more about in the "What Do I Need to Know?" column.

What Do I Know?	What Do I Need to Know?

3. Put a check mark by **one to three** questions or issues in the "What Do I Need to Know?" list that you think are most important to explore.

4. **What kinds of references or resources would help you answer or explore these questions?** Identify two different resources and explain what information each resource is likely to give that will help you answer the question(s). Choose specific resources.

Core Investigations

I. Critical Reading

To complete this investigation, you should have already read Concepts 5.1–5.3 in Chapter 5. In Concept 8.4, you should also read the text under the headings "Substrate Specificity of Enzymes" and "Effects of Local Conditions on Enzyme Activity." Then answer the following questions.

1. In the case narrative, Rob learned that the stain near the neckline of the dress contained starch. What specific types of macromolecule are starch and cellulose?

2. What monomer is found in starch and cellulose?

3. Contrast the structure and function of starch with those of cellulose in plant cells.

4. What is an enzyme?

5. To remove the stain from the dress, Rob treated the stain with a cleaner containing the hydrolytic enzyme *amylase*. Explain what happens to starch at the molecular level when it is acted upon by amylase. You may wish to sketch the structure of starch to show how this enzyme works (see Figure 5.7 in your text).

6. Under the right conditions, amylase breaks down amylose efficiently; however, the enzyme is not very effective in breaking down amylopectin. Examine Figure 5.6 and read the related text in your textbook. Use your observations to propose a hypothesis for why amylase breaks down amylose much more effectively than amylopectin.

7. Explain why Rob did not have to worry that the amylase cleaning solution would damage the dress.

II. Analyze and Design an Experiment

To further investigate starch and its components, first you will analyze an experiment. Then you will design your own. The experiment you will analyze was performed using the software in the *Chapter 41 Investigation: What Role Does Amylase Play in Digestion?* found on the Campbell website (http://www.masteringbio.com) and CD-ROM. However, you can complete the exercise with the information provided in this workbook.

A. Analyze an Experiment. In the following controlled experiment, we used both iodine solution (IKI) and Benedict's solution as indicators to test the effect of amylase on starch. As you may recall, Rob used the indicator iodine to test the dress stain for the presence of starch.

The Experiment: Four test tubes were set up. To find out which substances were placed in each tube, see the table in the bottom section of Figure 1.2. The tubes were then incubated at 37°C for 60 minutes (none were boiled). Half of the contents in tubes 1–4 were poured into tubes 1A–4A. The contents of tubes 1A–4A were tested with IKI. The remaining contents in tubes 1–4 were tested with Benedict's solution. The next set of questions asks you to analyze the results of both tests.

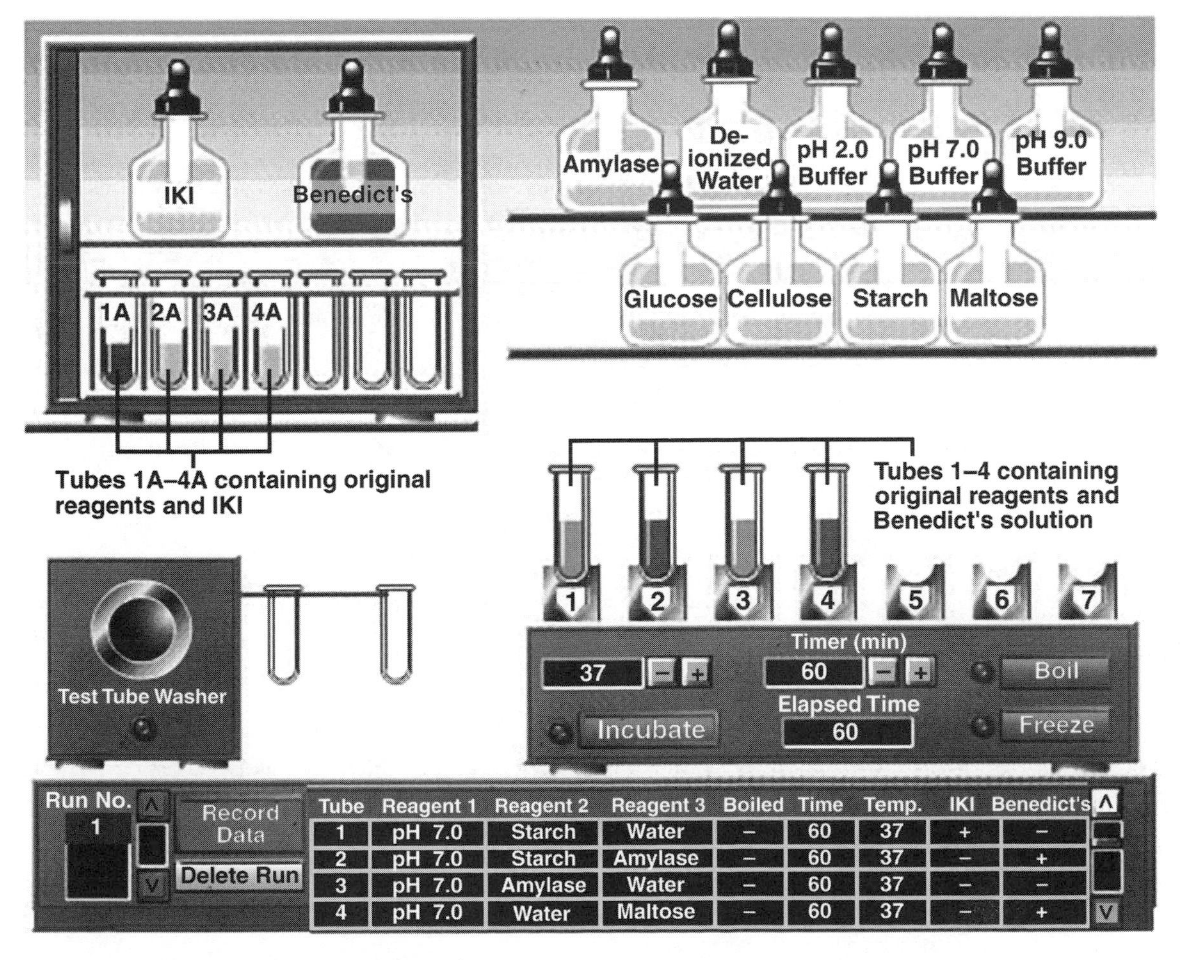

Tube	Reagent 1	Reagent 2	Reagent 3	Boiled	Time	Temp.	IKI	Benedict's
1	pH 7.0	Starch	Water	–	60	37	+	–
2	pH 7.0	Starch	Amylase	–	60	37	–	+
3	pH 7.0	Amylase	Water	–	60	37	–	–
4	pH 7.0	Water	Maltose	–	60	37	–	+

Figure 1.2 Experimental setup and results.

1. Review Figure 1.2 and note which of the reagents were used in each of the four test tubes. In Table 1.1, check off the reagents found in each tube.

Table 1.1 Test Tube Contents

Reagent	Test Tube			
	1	2	3	4
Starch				
Amylase				
Buffer pH 7.0				
Maltose				
Water				

2. Note the test tube results for the use of IKI. The color change in the contents of each test tube is also shown in the test-tube rack at the upper left corner of the figure. A dark shade indicates a positive iodine reaction (the actual color is dark blue) and a light shade indicates a negative reaction (the actual color is yellow). Because Rob used iodine in this case, you know that iodine is a test for starch. Why, then, was the iodine test negative for test tube 2A?

3. What is the purpose of adding iodine to test tubes 1A and 3A?

4. Now examine the results for the Benedict's solution test. Color changes are shown in the rack containing tubes 1–4 at the middle right. A dark shade indicates a positive test (the actual color is reddish brown). A light shade indicates a negative test (the actual color is blue). Do you think Benedict's solution is a test for starch, amylase, or maltose? Explain.

B. Design Your Own Experiment

1. Using *either* the iodine test or the Benedict's test, design an experiment to examine other factors in the action of amylase on starch. Your experiment should test only *one* of the following variables: pH, incubation temperature, incubation time, boiling, or freezing.

a. What question will you investigate in your experiment?

b. Restate the question as a hypothesis (see Chapter 1 in your textbook for an explanation of forming hypotheses):

If ____,

then ____.

2. In Table 1.2, describe your experimental treatment in test tube 1. Although it is likely that you will have more than one control tube, describe just one of your controls for test tube 2.

Table 1.2 Test Tube Contents and Conditions

Reagents	Test Tube 1 (experimental)	Test Tube 2 (control)
Starch		
Amylase		
pH		
Experimental conditions		
Incubation time		
Incubation temperature		
Boiling		
Freezing		

3. In Table 1.3, indicate the results you would expect if your hypothesis were supported (use "+" and "−" for the test indicator that you chose).

Table 1.3 Predicted Results, Supporting Hypothesis

Tests	Test Tube 1	Test Tube 2
Benedict's		
IKI		

4. In Table 1.4, indicate the results you would expect if your hypothesis were *not* supported (use "+" and "−" for the test indicator that you chose).

Table 1.4 Predicted Results, Not Supporting Hypothesis

Tests	Test Tube 1	Test Tube 2
Benedict's		
IKI		

5. *Optional.* Conduct the experiment you designed using the software provided in the *Chapter 41 Investigation: What Role Does Amylase Play in Digestion?* found on the Campbell website (http://www.masteringbio.com) or CD-ROM. Turn in a screen capture of the table showing your results. *Note:* Experiments involving IKI tests of cellulose will not give the correct results due to a bug in the software.

III. Off the Wall: Starch Degradation Investigation

Hildy planned to surprise her parents by remodeling their living room while they were away for the weekend. First she had to remove the wallpaper so that she could paint. When she started scraping at the edge of the dry wallpaper, only a few small pieces came off. "What's up with this wallpaper?" Hildy asked herself. "It's just not coming off!"

Hildy got a spray bottle and filled it with warm water. She sprayed the walls to moisten large areas. After several minutes, she scraped at the wallpaper again. Larger pieces came off this time, but big patches of hardened paste remained. Hildy couldn't spend the whole weekend scraping! She rummaged around the house and found some alcohol and some vinegar.

Unsure of what these substances would do to the walls, she also went out and bought two different types of commercial wallpaper remover. "I wonder which of these will work the best?" she thought.

To test which one would work best, she chose a section of the wall behind the couch and applied the five substances to a 10-cm^2 section of the wall. She labeled each patch to remember which substance had been applied to each square. After 20 minutes, she noted how much wallpaper she could remove with one scrape from each patch. See the results of her experiment in Figure 1.3.

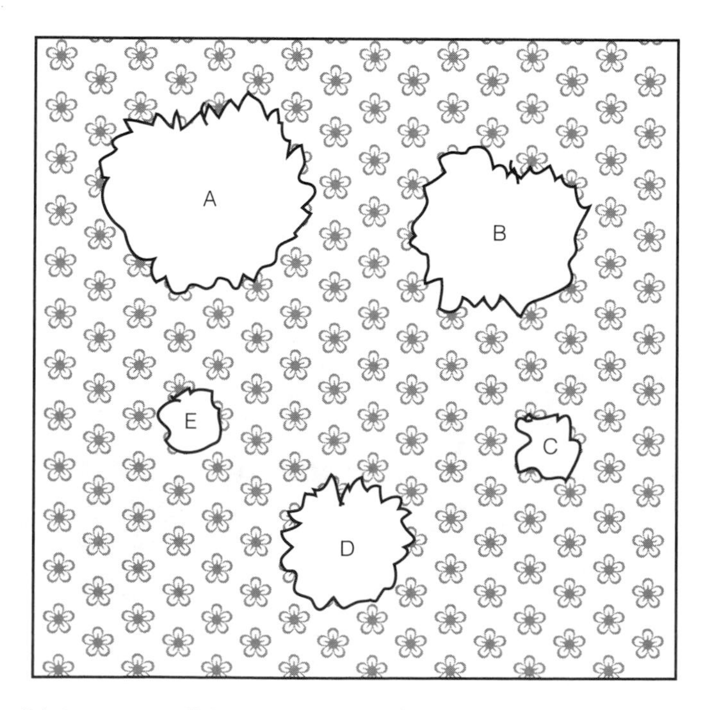

Figure 1.3 A section of Hildy's parents' living room wall after her experiment. The table on the next page is a key containing her results.

Label on Wallpaper	Substance	Approximate % of Wallpaper and Paste Removed
A	remover with 0.5% amylase	100
B	remover with 0.1% amylase	75
C	rubbing alcohol	10
D	vinegar	50
E	water	10

1. Which substance worked best? What does this tell you about the composition of wallpaper paste?

2. Describe how the most effective substance worked to remove the paste.

3. Considering that vinegar is an acid, explain the results seen with the vinegar.

4. Why was it important that Hildy also test the effect of water alone on the wallpaper paste?

Additional Investigations

IV. Structure and Function of Starches

A. Kinds of Starch. Starches are a significant part of the typical human diet, making up 40–80% of total energy intake. Some plants store more starch than others. Humans have discovered many varieties of starchy plants that satisfy our hunger and taste buds, such as corn, cassava, and potatoes, originally from South America; sweet potatoes and yams, from tropical Africa and South America; chickpeas, from Turkey; plantains, originally from India; rice, originally from Asia; soybeans, originally from China; and wheat, from the Middle East.

Plants store starch as highly condensed granules that do not dissolve easily in water. The composition and size of these granules vary in different types of plants.

1. Contrast the microscopic starch granules of corn with those of potato, shown in Figure 1.4.

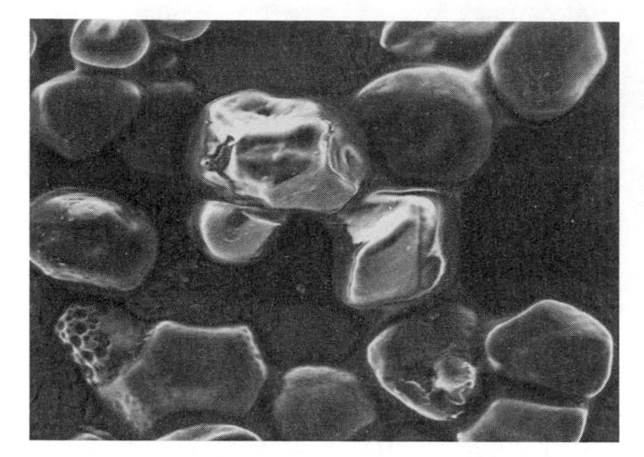

(a) Starch granules in corn (5–25 μm)

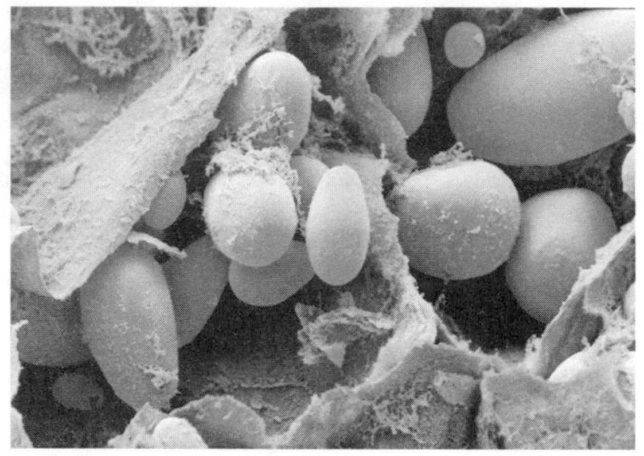

(b) Potato starch (15–100 μm)

Figure 1.4 Note the variations in the size and shape of starch granules.

2. Now that you have learned more about the different types of starch granules, can you infer the type of granules that Rob scraped from the old dress? Explain your response.

B. Using Starches in Food: Understanding Structure for Commercial Application. Although the enzymes in our digestive system are capable of breaking apart starch granules, cooking starchy foods causes the starches to gelatinize, which enhances texture and taste and improves digestion. *Gelatinization* is the process in which granules of starch swell, break up, and disperse in water. Suspensions of various thicknesses are formed during this process.

Figure 1.5 on the next page shows a cornstarch granule before and after gelatinization. (Cornstarch is often used as a thickening agent in food products such as gravies and sauces.) Starch granules have complex structures. The granule surface consists of many amylopectin and some amylose molecules associating tightly with each other due to hydrogen bonding. Water does not easily penetrate the granule. Tiny channels lead from the surface into an amorphous center where less tightly bound amylose and amylopectin molecules are found.

When starch granules are immersed in water, two things happen. Water moves slowly through the channels and forms hydrogen bonds with the amylose and amylopectin components in the center of the granules. At the same time, hydrogen bonding among the amylose and amylopectin molecules on adjacent granules causes clumping. Stirring the suspension prevents the granules from forming dense clumps. If understirred, the cornstarch mixture will be lumpy.

When exposed to heat, water molecules move more rapidly. The rapid movement allows more water molecules to enter the granules, causing the granules in the suspension to swell. The cornstarch and water mixture noticeably thickens (increases in viscosity). Because amylose molecules are unbranched, they can easily move through the channels and will leach out of the granules more quickly than amylopectin. If the gel sits and cools at this stage, the amylose molecules will begin to realign by hydrogen bonding, causing the granules to adhere to each other and to the container. The cornstarch may thicken unevenly and the resulting mixture will be difficult to pour.

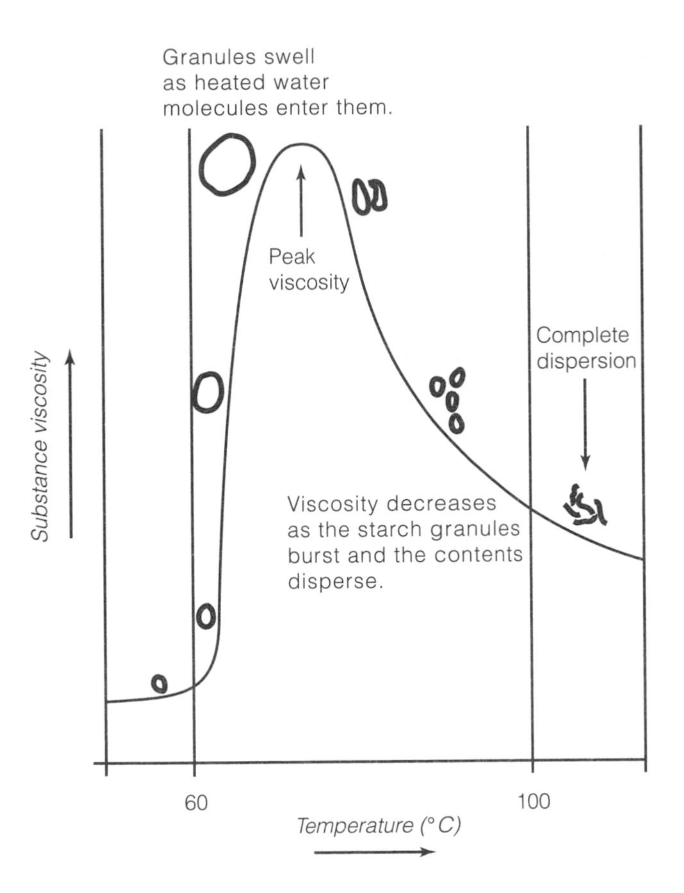

Figure 1.5 Gelatinization and disruption of starch molecules.

On the other hand, continued heating of the gelatinized starch transfers more energy to the water molecules, allowing them to further destabilize hydrogen bonds between starch molecules. The granules continue to swell, and more starch molecules leak into the surrounding liquid. Heating past the boiling point causes swollen granules to break into fragments and release all of the starch molecules into the water. At this point, the mixture thins (decreases in viscosity). Stirring the mixture will hasten the thinning process, leading to a runny cornstarch mixture.

Investigation: What Went Wrong?

Gravy and mashed potatoes are two foods prepared in many U.S. homes, but they are tricky to make successfully. Examine the recipes on the next page for gravy and mashed potatoes and answer the following questions using your knowledge of the gelatinization process. (*Hint:* The secret to making gravy and mashed potatoes is maintaining an even distribution of granular structure without fragmenting the individual granules.)

Turkey Gravy	Mashed Potatoes
1. In a large saucepan, over medium heat, bring 1 cup of turkey broth and pan juices to a boil. 2. Meanwhile, blend until smooth 2 tablespoons of cornstarch in 1 cup of cold water. 3. Slowly add the cornstarch mixture to the boiling broth. 4. Stir intermittently until the gravy thickens. 5. Season to taste with salt and pepper. 6. Remove from heat and serve immediately.	1. Place a large pot of cold water on the stove. 2. Peel each potato, cut into cubes of about 3/4-inch square, and put in the pot. 3. Do not allow the water to boil too rapidly; check for doneness after 15 minutes. 4. Mash the potatoes while they are hot. Do not overmash. Never use a food processor. 5. Mix in butter and milk. Do not let potatoes cool.

1. As you begin to prepare turkey gravy, you carefully blend 2 tablespoons of cornstarch into a cup of cold water. You add the mixture to the turkey broth, but then you forget to stir it. Your gravy turns out lumpy. How did the mistake ruin the gravy?

2. To make mashed potatoes, you boil the potatoes for 25 minutes. The potato chunks begin to disintegrate as you drain them. When you add the butter and milk, the potatoes are thin and gluey—similar to wallpaper paste before it dries. What went wrong?

C. Structural Properties of Native and Modified Starches in Commercial Products. Starch has a number of properties that make it useful to manufacturers of prepared foods and other commercial products such as glues. Cornstarch products, such as corn syrup, are among the most common ingredients listed on the food labels of cookies, puddings, frozen dinners, and crackers.

Naturally occurring starches (native starches) may be used in dry form as ingredients for foods (about 75% of wheat flour is starch) or as dry lubricants (baby powder), but most are added to water to create gels and solutions. Two widely used types of native cornstarch, *dent* and *waxy,* vary in their amylose content, which makes them useful for different purposes.

Dent cornstarch comes from the most frequently planted type of corn in the United States. Dent cornstarch usually contains about 80% amylopectin and 20% amylose. Starch products made with dent corn tend to adhere to surfaces and form more rigid layers as they are cooked or allowed to dry. For example, dent cornstarch is used in the production of wallpaper paste. Amylose causes the wallpaper to stick to the wall through hydrogen bonding with cellulose and then to stiffen as it dries. The harder outer coating of jelly beans is also made from dent starch.

Waxy cornstarch is produced by a type of corn plant that does not produce amylose. Waxy starch consists entirely of amylopectin molecules. When this cornstarch is dissolved in a solution, it tends to be more stable than the dent cornstarch. The resulting product pours easily. For example, hot chocolate mixes contain waxy cornstarch.

Manufacturers often chemically modify native cornstarch to form additional bonds that cross-link amylose molecules or cross-link amylose and amylopectin molecules. These modified starches have different chemical properties than the native starches.

- Cross-linked waxy starches like Consista® and Rezista® absorb water but retain their granular structure, producing more stable mixtures with higher viscosity than that found in native starches. Products requiring a thicker consistency, such as gravy in canned stew, often contain modified waxy starch.
- Amylomaize, another modified starch, contains 70% amylose and 30% amylopectin. Manufacturers use amylomaize to make inexpensive and biodegradable packaging foam with good cushioning and resiliency properties. For starch to act like polystyrene (a plastic), its polymer molecules have to align closely through hydrogen bonds. Linear molecules perform better in this way than branched molecules; therefore, the high amylose content of amylomaize makes it work well.

Read the following product description and determine which starch would be the best choice for a manufacturer. (*Note:* The cost of dent starch is low, waxy starch is more expensive, and chemically modified waxy starches are the most expensive. Although cost is always an important factor in manufacturing decisions, for this exercise consider only the characteristics of the different types of starch.)

A. Dent starch
B. Waxy starch
C. Modified waxy starch
D. Amylomaize

1. **Instant cheesecake mix.** Manufacturers need a starch that will maintain a creamy consistency and will neither liquefy nor harden at room temperature. Explain your choice.

2. **Soups.** Manufacturers need a starch that allows their product to be pourable but does not thicken too much as it cools. Explain your choice.

3. **Batter and breading.** Manufacturers need a starch that will adhere to chicken and then become crunchy as the chicken is cooked. Explain your choice.

V. Open-Ended Investigations

Why is starch used in papermaking? Consider the structure of starch molecules in your answer.

References

Stanley, Keith D., Senior Research Scientist, Tate & Lyle, Decatur, IL. Personal communication.

Tate & Lyle (manufacturer of carbohydrate ingredients) http://www.tateandlyle.com/TateAndLyle/products_applications/product_application_grids/americas/default.htm (accessed June 29, 2007).

Whistler, R. L., and E. F. Paschall. *Starch: Chemistry and Technology.* New York: Academic Press, 1965.

CHAPTER 2

Bean Brew

Bean Brew

Henry, Edie, Taki, and Sally sat around the table at their favorite restaurant celebrating Henry's new job. "I can't believe it's already been six years since we met," Sally said.

It wasn't long before the talk around the table turned to biotechnology stocks. Edie and Taki were always well informed about the latest companies and enjoyed arguing about what products were going to be the "next big thing."

"Excuse me," Sally began with a smile when there was a break in the animated conversation. "What's all the fuss about a new strain of transgenic fungus? I can't imagine how this would affect me . . ."

Taki reached for the small container of soy sauce on the table and held it up. "It turns out that this fungus will increase the efficiency of the first stage of brewing soy sauce. Did you know that brewing soy sauce is one of the original biotech industries? They were shipping the stuff in barrels in Asia over 500 years ago and in bottles to Europe by the 1600s. Now most of the world uses soy sauce."

The friends settled in; considering Taki's usual attention to detail, this would be a long story. "About 5,000 years ago in China," he began, "people grew soybean crops for food and animal feed. Storing beans was risky because of spoilage. Salt was added as a preservative, but over time the beans fermented."

"Like pickles and sauerkraut?" Henry asked.

Taki nodded and continued, "Except the beans softened as they fermented. This paste was easier to digest, so people started to eat it. It's called miso today. Then, about 500 years ago, someone discovered that the liquid in the bottom of the barrel could be used for cooking. And so, soy sauce was invented!"

"Is this fermentation process similar to making wine from grapes?" Sally asked.

"Well, soy sauce brewing is actually done in two stages. In Japan soy sauce is called *shoyu.* To make it, you first steam the soybeans and mix them with toasted, crushed wheat. Then add the fungi *Aspergillus oryzae* and *Aspergillus sojae.* The new mixture, called *koji,* is left uncovered for a couple of days, while the fungi partially digest the soy and wheat."

"So, is the transgenic fungus you were talking about *Aspergillus*?" asked Sally.

"Exactly," Taki replied. "Okay, in the next stage, you mix the koji with water and a lot of salt to form a mash called *moromi*. Then put the moromi into airtight containers and let them ferment for at least 6 months. Squeeze this mash to get the liquid soy sauce, which is filtered, pasteurized, and tightly bottled. So that's it—soybeans, wheat, water, salt, and microbes. Back in the days of the empire, they even had special recipes that they made only for the emperor by adding extra flavors."

"So what kind do we have here?" asked Edie.

"Oh, an emperor's brew, for sure," asserted Henry. They all laughed.

Figure 2.1 Commercial production of soy sauce involves the extended fermentation of moromi, a mash of soybeans and wheat in a salt solution.

CASE ANALYSIS

1. **Recognize potential issues and major topics in the case.** What is this case about? Underline terms or phrases that seem to be important to understanding this case. Then list **three or four** biology-related topics or issues in the case.

2. **What specific questions do you have about these topics?** By yourself, or better yet, in a group, list what you already know about this case in the "What Do I Know?" column. List questions you would like to learn more about in the "What Do I Need to Know?" column.

What Do I Know?	What Do I Need to Know?

3. Put a check mark by **one to three** questions or issues in the "What Do I Need to Know?" list that you think are most important to explore.

4. **What kinds of references or resources would help you answer or explore these questions?** Identify two different resources and explain what information each resource is likely to give that will help you answer the question(s). Choose specific resources.

Core Investigations

I. Critical Reading

To complete this investigation, you should have already read Chapter 7: Membrane Structure and Function (specifically Concepts 7.1 and 7.3); Chapter 8: An Introduction to Metabolism (specifically Concept 8.4); and Chapter 9: Cellular Respiration.

A. The Koji Phase. In the koji phase of soy sauce production, fungi produce enzymes that break down the carbohydrate and protein in the soybeans and wheat, thereby obtaining energy and molecules for fungal growth. Recall that koji is left uncovered for a few days, which allows many other types of microbes to enter the soybean-and-wheat mixture.

1. Describe a typical enzyme-substrate complex. What mechanisms do enzymes use to lower activation energy and speed up a reaction?

2. Explain how enzymes break down macromolecules. What is the role of water? What bonds are broken, what bonds are formed? Examine Figure 8.17 as you develop your answer.

3. In the koji stage of soy sauce production, *Aspergillus* fungi digest soybeans and wheat. *Aspergillus* uses some of the glucose produced by the breakdown of the carbohydrates to generate ATP through cellular respiration or fermentation. Examine Figure 9.9 and answer the following questions about glycolysis, the first stage of respiration.
   a. How many different enzymes shown in Figure 9.9 are used to transform glucose into pyruvate?

b. What types of reactions do isomerases catalyze?

c. What kinds of enzymes catalyze reactions that transfer a phosphate group from ATP to another molecule?

d. If you added an aldolase inhibitor, what key reaction would be unlikely to occur? Explain.

B. The Moromi Phase. Once *Aspergillus* has broken down the macromolecules in the soybeans and wheat into monomers, the koji phase ends. Moromi is then made by mixing the koji with water and enough salt to make a 16–20% concentrated salt solution, or brine.

1. In the moromi phase of soy sauce production, the osmotic conditions for microbes are drastically changed. Sketch a generic cell showing what happens to most cells when they are placed in brine. Explain your sketch. (Hint: Consider the movement of water.)

2. Some microbes have adaptations for osmoregulation in order to live successfully in high-salt environments. When the brine is added, the populations of bacteria and fungi found in the koji change. Do you expect greater or lesser microbial diversity? Why?

3. Yet another challenge faces the microbes in moromi. After the brine is added, workers place the moromi in airtight containers for several months. Which types of microbes will survive under these conditions? Explain how they will obtain energy for life processes.

4. *Tetragenococcus halophilus* (a bacterium) and *Zygosaccharomyces rouxii* (a fungus) are two facultatively anaerobic species that thrive in moromi. Through fermentation, *Tetragenococcus* produces lactic acid (lactate in its ionized form) and *Zygosaccharomyces* produces ethanol. What molecule is transformed into these waste products? Describe the two processes. What other waste products are produced?

5. Are ethanol and lactate oxidized or reduced in these reactions?

II. Fermentation of Grapes

A. Yeast and Rising Alcohol Concentrations. One of the oldest uses of fermentation by people is to make alcoholic beverages such as wine. However, fermentation also occurs without human intervention. Once grapes ripen on the vine, tiny breaks in the skin of the fruit enable the entry of microbes such as bacteria and fungi. The interior of the grape provides both a high concentration of sugars and low pH. Fermentative yeasts thrive in this environment and metabolize the grape sugars for energy. The products carbon dioxide and ethanol are rapidly transported out of the cells as wastes.

When people make wine by fermenting grapes, the process occurs within an airtight container. Alcohol continues to build up in the container until the alcohol tolerance level of the specific yeast population is reached, ending the fermentation cycle. Figure 2.2 shows the results from a simulation of wine fermentation over a 10-day period.

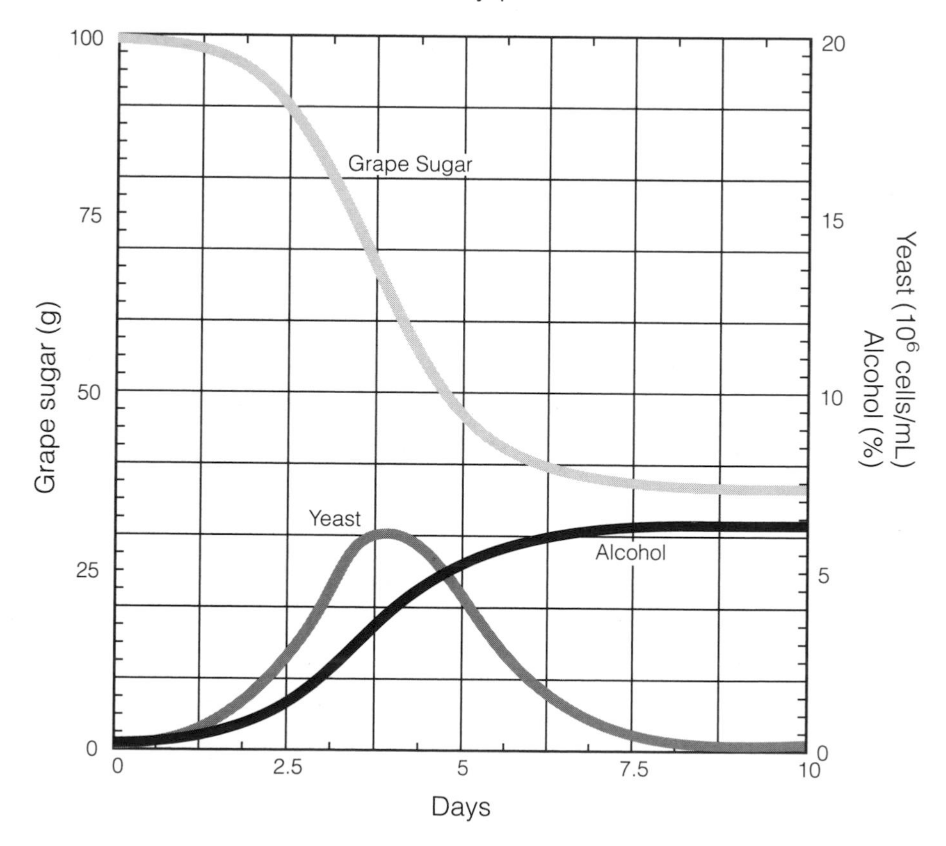

Figure 2.2 Results from a simulation of wine fermentation (Stanley et al., 2003). The graph shows changes in grape sugar, yeast population, and percentage alcohol over a 10-day period. (*Note:* Read grape sugar on the left axis. Yeast and alcohol are shown on the right axis.)

1. Examine Figure 2.2 and fill in the information below.
   a. The grape sugar level starts at ______ g and ends at ______ g.
   b. The yeast population reaches its highest level of approximately ______ on Day ______.
   c. The alcohol level starts at ______% and ends at approximately ______%.
   d. Look at the graphs showing the correlation between yeast population and percentage alcohol. At what percentage alcohol does this yeast population begin to decline? ______%

2. Why isn't the remaining grape sugar converted to ethanol and carbon dioxide?

3. What product of alcohol fermentation is not shown in the graph in Figure 2.2?

4. If you removed the alcohol as it was produced, would you predict an increase or a decrease in the amount of grape sugars at 10 days? An increase or decrease in the population of yeast at 10 days? Explain.

5. A bottle of wine may spoil if it is allowed to sit for some time after being opened or if its cork does not form a tight seal. Explain what causes the wine to spoil under these conditions. (*Hint:* Available grape sugar declines.)

B. Fermentation with Wild and Cultivated Yeasts. In an experiment to identify differences in fermentation carried out by wild and cultivated yeasts, a batch of grapes was divided in two. One batch of grapes was treated with sulfur dioxide to kill wild yeasts before the juice was extracted. The other batch was left untreated, allowing wild yeasts to survive.

Fermentation of grape juice extracted from these two groups was carried out in separate containers. In the first container, the juice from the treated grapes was inoculated with a special cultivated strain of yeast. The untreated juice in the second container was inoculated with only wild yeast populations. Both containers were allowed to ferment for 10 days. Samples were removed daily to estimate the number of yeast cells and the level of alcohol in each container. Results are shown in Figure 2.3.

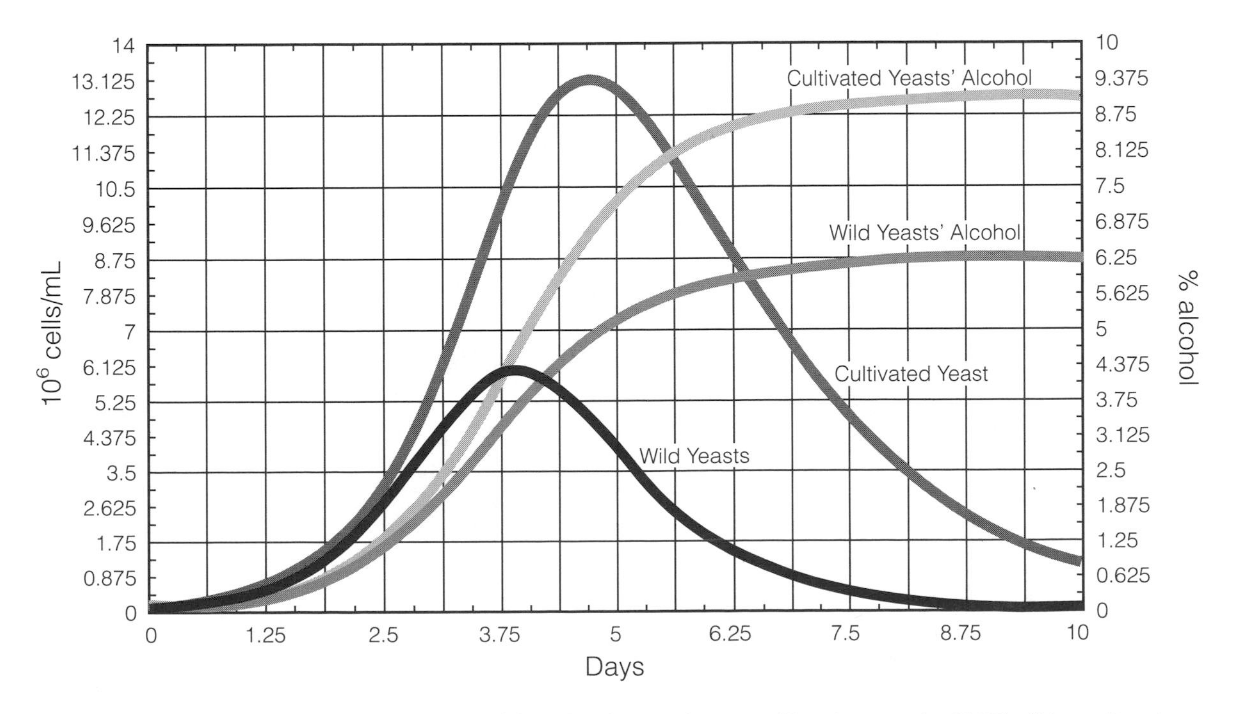

Figure 2.3 Simulated fermentation by wild and cultivated yeasts (Stanley et al., 2003). (*Note:* Read population size on the left axis. Alcohol production is shown on the right axis.)

1. Assuming alcohol level affects the growth of yeast, which yeast has a higher tolerance for alcohol? At approximately what percentage alcohol do the two yeast populations in the different containers begin to decline?

2. Why do you think the alcohol levels increase more rapidly in one of the containers? Use data from Day 3.75 to support your hypothesis.

C. Bottling Soy Sauce. Now apply some of the concepts you learned about grape fermentation to the Bean Brew case. When soy sauce was first shipped to Europe, Asian soy sauce producers tried the same method they had used for shipping shorter distances within Asia—simply filtering the soy sauce and placing it in non-airtight containers. However, the soy sauce always spoiled before it reached its European destinations! The spoilage problem was solved when the producers started to boil the soy sauce first and then place it in airtight bottles.

1. Explain why placing soy sauce in airtight bottles was more successful for long-distance shipping than simply placing the sauce in barrels.

2. When the soy sauce was not boiled before it was bottled, the bottles sometimes burst during the voyage. What do you think caused this?

3. Bottled soy sauce does not taste the same as fresh soy sauce. What do you think causes this change?

4. To preserve flavor in modern times, brewed soy sauces are not boiled but are pasteurized (heated to a temperature of about 60°C [140°F]) before being bottled. Pasteurized soy sauce tastes better than boiled soy sauce. What does pasteurization do? Why should opened bottles of soy sauce be stored in the refrigerator?

III. Alcohol Dehydrogenase

Ethanol, which is toxic to yeast cells, is also toxic to human cells. We can consume alcohol due to alcohol dehydrogenase, an enzyme produced by humans and many other animals that catalyzes the oxidation of alcohols to aldehydes. In this reaction, nicotinamide adenine dinucleotide (NAD^+) is used as an oxidizing agent.

$$\underset{\text{ethanol}}{CH_3CH_2OH} + NAD^+ \xrightarrow{\text{alcohol dehydrogenase}} \underset{\text{acetaldehyde}}{CH_3CHO} + NADH + H^+$$

Not only does alcohol dehydrogenase allow humans to detoxify (within limits!) the ethanol that we consume, but also it detoxifies the alcohol produced by certain fermentative microbes that reside in our small intestine and colon.

1. Draw molecules of ethanol and acetaldehyde.

Ethanol Acetaldehyde

2. Explain why ethanol is considered an electron donor in this redox reaction.

3. Consumption of methanol can be fatal because alcohol dehydrogenase converts methanol to formaldehyde, a highly toxic substance that can cause the death of cells in the human body. Formaldehyde is the substance once commonly used to preserve animal and plant tissues; however, due to its cancer-causing properties, its use is restricted. Draw molecules of methanol and formaldehyde.

Methanol Formaldehyde

4. What is oxidized and reduced in this reaction? Explain.

$$CH_3OH + NAD^+ \xrightarrow{\text{alcohol dehydrogenase}} CH_2O + NADH + H^+$$

methanol → formaldehyde

5. Treatment for methanol ingestion involves giving the patient an alcohol dehydrogenase inhibitor. Explain why this is helpful.

6. During fermentation in yeast, alcohol dehydrogenase catalyzes a reaction that breaks down acetaldehyde into ethanol and regenerates NAD^+. Note that this is the reverse of the reaction catalyzed by the enzyme in humans and other animals.

$$CH_3CHO + NADH + H^+ \xrightarrow{\text{alcohol dehydrogenase}} CH_3CH_2OH + NAD^+$$

acetaldehyde → ethanol

a. What is oxidized and reduced in this reaction? Explain.

b. What happens to the ethanol after it is produced?

c. What happens to the NAD^+ after it is produced?

Additional Investigation

IV. More Human Uses of Fermentation

Things are fermenting everywhere! Choose a product from the list below. Use your text and other resources, including primary sources, to find out how fermentation is used to make this product. Write a paper of one to three pages, based on reliable sources, indicating:

- the organism(s) doing the fermenting
- the metabolic pathway(s) used
- substrates
- fermentation products
- how the fermentation is accomplished
- how the product is prepared for consumption

Products

Sausage
Chocolate
Coffee
Sourdough bread
Cheeses
Tempeh
Kimchee
Sauerkraut
Citric acid (widely used as an ingredient)
Dental caries (product is the decayed tooth)
Vinegar
Yogurt

V. Open-Ended Investigations

Use the working wine model (available at http://bioquest.org/icbl/casebook/wine) to conduct your own investigations of factors involved in wine fermentation.

Additional pairs of graphs (A+B or C+D) are available on the same website for further practice in interpreting graphs, making inferences, and drawing conclusions.

References

Noda, F., K. Hayashi, and T. Mizunuma. Antagonism between osmophilic lactic acid bacteria and yeasts in brine fermentation of soy sauce. *Applied and Environmental Microbiology,* 40(3):4452–457, 1980.

Stanley, Ethel D., Howard T. Odum, Elisabeth C. Odum, and Virginia G. Vaughan. Modeling wine fermentation, pp. 85–92, and software on CD-ROM. In J. R. Jungck, M. F. Fass, and E. D. Stanley, *Microbes Count!* Beloit, WI: BioQUEST Curriculum Consortium and American Society for Microbiology Press, 2003.

CHAPTER 3

The Donor's Dilemma

The Donor's Dilemma

Usually, Russell found an excuse not to participate in company-sponsored blood drives, but for the first time he decided to donate blood. After filling out the donor eligibility form and passing the blood pressure, pulse, temperature, and blood-clotting tests, Russell sat down for his interview.

Russell interrupted the long list of "Have you ever?" questions with a question of his own. "What if I have West Nile virus?"

"West Nile virus is uncommon," the interviewer said. "Besides, all donated blood is tested for West Nile virus, even here in California where it's extremely rare." She glanced over his paperwork. "Let's see. You said you haven't had any fevers or headaches in the last week. Is there a reason that you think you might have it?"

"No, but I've heard that sometimes people don't have any symptoms," Russell responded. "I just got back from a hiking trip in Boulder, Colorado, over the Fourth of July weekend. There were news reports about a lot of cases of the virus there, and I'm still covered with mosquito bites."

"Well, if you have West Nile virus, we will find out. Lab tests on your blood will identify the presence of genetic material from the virus," the interviewer said reassuringly. "WNV can only be transmitted through blood transfusions if there are virus particles in the donated blood. In the United States, only a tiny fraction of blood donations last year tested positive for West Nile virus."

"So if I have West Nile virus, could you tell if I got it in Colorado?" Russell asked.

"Well, they can't tell from this blood screening, but other tests can identify the strain of WNV," she replied. "When West Nile virus first appeared in New York in 1999, all the samples were alike. But now mutations are showing up in the virus as it migrates to different areas. We're seeing different strains of the virus in different regions of the country."

"So did West Nile virus originate in New York?" Russell wondered.

"No," she said with a smile, "it's called *West Nile* for a reason."

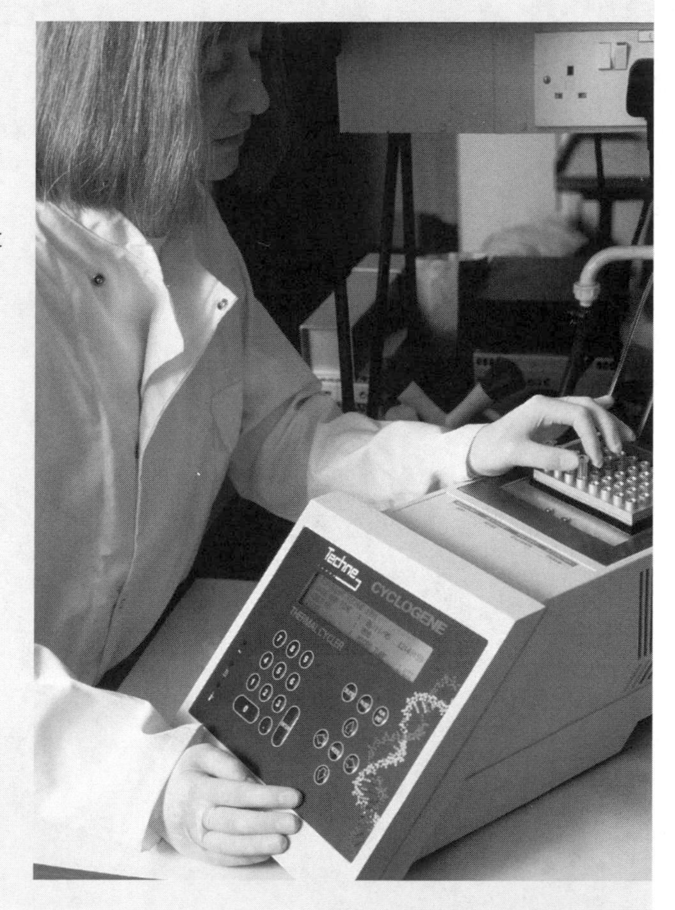

Figure 3.1 A lab technician performs a PCR, which is used to make multiple copies of DNA. Blood donations are tested for WNV using a form of PCR.

CASE ANALYSIS

1. **Recognize potential issues and major topics in the case.** What is this case about? Underline terms or phrases that seem to be important to understanding this case. Then list **three to four** biology-related topics or issues in the case.

2. **What specific questions do you have about these topics?** By yourself, or better yet, in a group, list what you already know about this case in the "What Do I Know?" column. List questions you would like to learn more about in the "What Do I Need to Know?" column.

What Do I Know?	What Do I Need to Know?

3. Put a check mark by **one to three** questions or issues in the "What Do I Need to Know?" list that you think are most important to explore.

4. **What kinds of references or resources would help you answer or explore these questions?** Identify two different resources and explain what information each resource is likely to give that will help you answer the question(s). Choose specific resources.

Core Investigations

I. Transmission of West Nile Virus (WNV)

West Nile virus is an arbovirus (**ar**thropod **bo**rne) that infects birds, humans, and other animals. Although the virus was first detected in Uganda in 1937, the first bird and human cases of West Nile virus in the United States were reported in New York City in 1999. Since then it has spread throughout much of North America. Mosquitoes are the vectors for the virus, transmitting it to the animals that they feed upon. Although mosquitoes feed on many types of vertebrates, birds are the most likely source of the virus. The virus multiplies at a very fast rate in the blood of many bird species, producing a high viral titer (concentration of virus particles in blood). Many bird species are known as *reservoir hosts* for the virus because they can "store" a high concentration of virus particles in their blood. When a mosquito feeds on the blood of a reservoir host, it will likely take in enough virus particles to transmit the virus to another potential host (Figure 3.2).

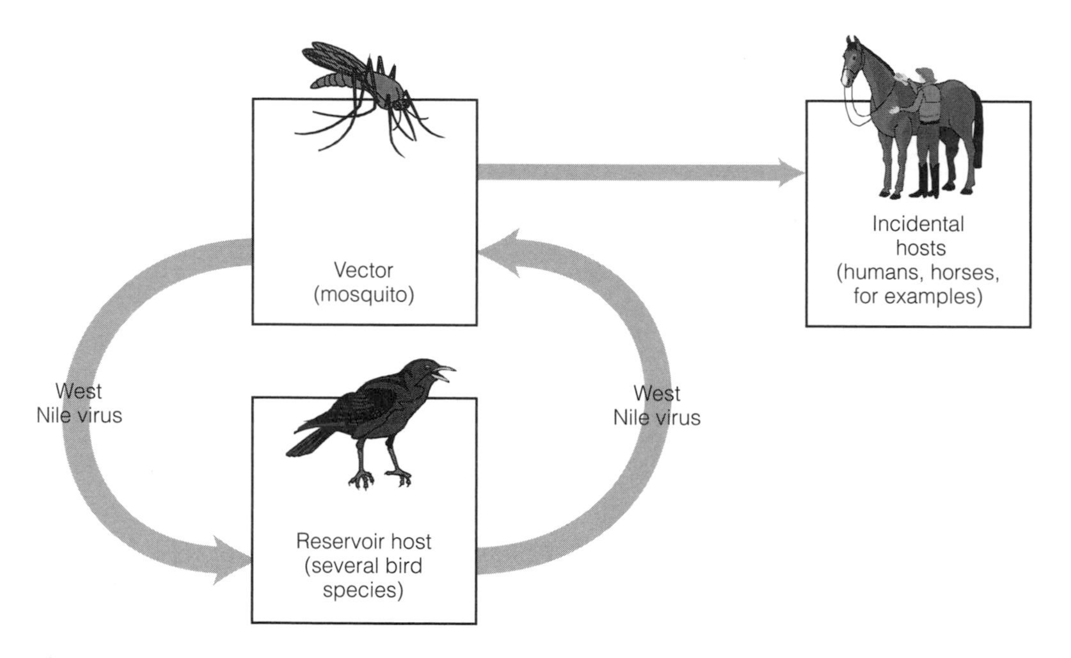

Figure 3.2 West Nile virus transmission cycle.

The interactions between infected birds and mosquitoes can quickly increase the incidence of WNV in a particular location, resulting in a cycle of viral amplification. The more mosquitoes there are, the more the virus is spread. The more birds that are present to be infected, the greater the number of virus particles that will be available to more mosquitoes.

Other animals, including humans, often serve as *incidental hosts*. Within incidental hosts, WNV is less efficient at multiplying; therefore, the concentration of the virus in the blood of these animals during infection is too low for mosquito vectors to pick up and transmit WNV to another host. Typically, these animals do not contribute to the cycle of amplification. West Nile virus can also be transmitted when an organism eats an infected organism. For example, crows that feed on the decaying flesh of other birds may contract West Nile virus through bird-to-bird transmission.

Eighty percent of humans who are infected with WNV show no symptoms. Twenty percent of those infected may experience fever, headache, fatigue, and body aches. Of the 20% with symptoms, only about 1 person in 150 develops encephalitis, a serious swelling of the brain that can cause death. In cases of WNV, viremia—presence of virus in the blood—lasts approximately 6 days or less. The amount of genetic material produced from the virus (a measure of viral titer) averages less than 5,000 copies of the virus per milliliter of blood. By comparison, other forms of viral encephalitis can result in a titer of 25,000,000 copies per milliliter of blood.

Human-to-human transmission of WNV through blood and organ donation, as well as during pregnancy or nursing, has been reported. Screening tests, including nucleic acid amplification and antibody detection, have been developed for WNV. (You will learn more about nucleic acid amplification in Investigation II.) Antibody detection tests are not used for screening blood because by the time the immune system produces antibodies in detectable amounts, the majority of the virus particles have been destroyed.

1. Several alligator farms in the southeastern United States reported an unusually high number of alligator deaths between 2001 and 2003. WNV was determined to be responsible for many of these deaths. Blood samples from infected alligators revealed high titers (some of which were higher than the titers in reservoir host bird species) for WNV. Considering that an adult alligator's hide is too thick for mosquitoes to penetrate (except for a few areas of soft tissue, such as inside the mouth and around the eyes), what are some other ways in which the alligators might have acquired WNV?

2. How would you add alligators to the transmission cycle shown in Figure 3.2?

3. Although humans produce low titers of WNV particles in their blood and don't serve as reservoirs for this vector-disseminated disease, human-to-human transmission of WNV is possible. Explain how a transfusion of infected blood can result in the dissemination of WNV.

II. Critical Reading

Before delving further into this investigative case, you first should read Concepts 17.1, 17.2, and 17.4, "Types of Point Mutations" in Concept 17.5, and Concepts 19.1 and 19.3. You might also want to do two Chapter 17 Activities on the Campbell website (http://www.masteringbio.com) or CD-ROM—*Overview of Protein Synthesis* and *Translation*.

In "The Donor's Dilemma," Russell wondered if it would be possible to tell where someone contracted West Nile virus. This is indeed possible. West Nile virus is an RNA virus. Like other RNA viruses, it has a high mutation rate; therefore, the nucleic acid sequence of a virus strain in New York could be quite different from a virus strain found in Egypt, for example. Many strains of WNV have been identified, and information about their nucleic acid sequences are stored in publicly available databases such as GenBank.

The sequences found in these databases are actually DNA sequences. In a laboratory, it is possible to create a DNA version of an RNA genome by using enzymes called reverse transcriptases. The newly constructed DNA sequence can be compared quickly to the sequences stored in databases by using powerful software to perform the comparisons. In the following activity, you will manually compare a short sequence of DNA (50 nucleotides out of 11,000) from six samples of WNV collected in Africa and Europe (Table 3.2). This particular sequence is part of the gene that codes for a portion of the virus's envelope protein (*E* gene).

1. Before you begin your analysis of the nucleotide sequences, use the data in Table 3.1 to make a prediction about the sequence that you would expect to be most similar to the one from Egypt. Make a second prediction about the one you would expect to be most dissimilar. Include number, country, and year.

Table 3.1 Identification of DNA Samples for a Portion of the Envelope (*E*) Gene of WNV (Berthet et al., 1997)

No.	Country	Year
1	Egypt	1951
2	France	1965
3	Senegal	1979
4	Senegal	1990
5	Uganda	?*
6	Madagascar	1986

*The specific year in which this sample was gathered in Uganda is unknown; however, it was after 1951.

Most similar:

Reason:

Most dissimilar:

Reason:

2. To analyze the sequences in Table 3.2 (see the next page), you will use manual methods that were used by geneticists until the development of computer-based methods. However, to make your comparison easier, a software program has been used to align the sequences in the table. The basic technique for comparing sequences has three steps:

   Examining the sequences for noticeable differences in length

   Comparing the sequences nucleotide by nucleotide

   Translating the sequences from codon to amino acid

   a. Consider Sequence 1, the oldest sequence from the West Nile region of Egypt, to be the standard for comparison. Examine the sequences shown in Table 3.2 for noticeable differences in length. Gaps in sequences are sometimes inserted by the computer as it aligns the rest of the sequence. These gaps are not present in the actual nucleic acid; however, they show up in the computer's output and often indicate certain kinds of mutations. Which of the sequences has either a deletion (gaps leading to a shorter length) or an insertion (leading to longer length)? Which type of mutation is it? Indicate, by column number, the affected nucleotides.

Table 3.2 Alignment of Six Sequences of Part of a WNV Gene for Envelope Protein*

```
  1       10        20        30        40        50
1 CCAACCACTGTGGAGTCGCATGGAAACTACTCCACACAGATTGGGGCCAC

2 CCAACCACTGTGGAGTCGCATGGAAACTACTCCACACAGATTGGGGCCAC

3 CCGACGACCGTTGAATCTCATGGCA            AGATAGGGGCCAC

4 CCAACCACTGTGGAGTCGCATGGAAACTACCCCACACAGATTGGGGCCAC

5 CCAACGACCGTTGAATCTCATGGCAGTTATTCAGCACAGATAGGGGCCAC

6 CCGACGACTGTTGAATCTCATGGCAATTATTCAACACAGGTTGGGGCCAC
```

*(Note that published DNA sequences, such as those shown here, are always the nontemplate strand of DNA; thus, it is directly comparable to mRNA. By replacing the *T*s with *U*s, these sequences can be directly translated using Figure 17.5 in your text. These are only fragments of the *E* gene sequence shown with the 5′ end to the left. The WNV genome is an open reading frame that starts before these first 50 nucleotides of the *E* gene.)

b. Next, analyze the differences in the columns of nucleotides to identify point mutations. Use a straightedge to keep your place, a highlighter, and a pen. Examine each vertical column in Table 3.2 starting at the left to look for variations from Sequence 1. If the nucleotides in a column match those of the standard sequence, highlight them. If there are deviations from the standard, circle them with the pen. For example, the first column contains all *C*s, so the whole column should be highlighted. The third column has two *G*s that vary from the A in Sequence 1. The two *G*s would be circled in pen and all the *A*s highlighted. How many total point mutations did you identify?

c. Determine the percentage of point mutations in sequences 2 through 6 (number of point mutations/number of nucleotides in sample × 100%). Sequence 2 is done for you as an example. (*Note:* For Sequence 3, count only the nucleotides present in the sequence.)

Sequence 2 = (0/50) × 100% = 0%

Sequence 3 =

Sequence 4 =

Sequence 5 =

Sequence 6 =

Which sample shows the greatest difference in nucleotides from Sequence 1? Explain. (*Note:* The 12 missing nucleotides in Sequence 3 should be considered as one deletion mutation rather than 12 point mutations because this deletion most likely occurred as one event.)

d. In the third and final step in comparing sequences, translate each of the six sequences from codon to amino acid, using Figure 17.5 in your textbook. Then you will be able to observe the consequences of the different mutations on the resulting polypeptides. Normally, you would expect to see a start codon (AUG), but assume instead that the reading frame begins with the first nucleotide at the 5′ end. Write in the appropriate amino acids under the DNA sequences in Table 3.2.

e. Examine each sequence. How many amino acids differed from the standard in sequences 2 through 6? Which amino acids changed?

Sequence 2:

Sequence 3:

Sequence 4:

Sequence 5:

Sequence 6:

What does this information reveal about the effects of the mutations on the *E* gene and the protein it codes for?

f. How many point mutations were involved in the amino acid differences you found? In Table 3.2, draw an asterisk by those nucleotides that made these differences.

g. How many of the point mutations were nonsense mutations? How many were silent mutations?

h. Compare your answers in 2c to those in 2f. Is the percentage of point mutations related to how many amino acids are changed? Explain your response.

i. Is it likely that the deletion mutation is also a frameshift mutation? Explain.

j. Now that you have identified, categorized, and determined the consequences of the various mutations in these sequences of WNV, how do these results compare to your predictions in question 1?

III. West Nile Virus: Viral Structure and Life Cycle

West Nile virus is a relatively small, spherical virus whose genome is single-stranded RNA (ssRNA), which also serves as the messenger RNA (mRNA) coding for viral proteins (Figure 3.3). This genetic material is contained within an inner protein coat called a capsid. Like many other animal viruses, WNV also has a membranous envelope derived from the host cell. This membrane surrounds the capsid and has numerous glycoproteins (the E protein) encoded by the viral genome. These glycoproteins are located on the outer surface of the envelope and function in the recognition of potential host cells. You analyzed the sequence of a portion of this viral envelope glycoprotein gene in Investigation II.

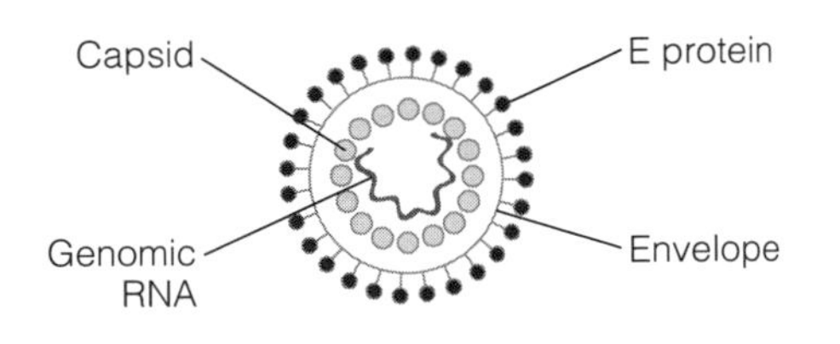

Figure 3.3 West Nile virus structure.

1. Animal viruses are classified by the type of nucleic acid found within the capsid. Using Table 19.1 in your textbook and the clues provided in the passage above, identify the classes for WNV and HIV. Provide an example of another virus from the same class for each.

   WNV Class ______ HIV Class ______

   Example: Example:

2. Compare the structure of WNV to that of HIV (see Figure 19.8 in your text).

3. How do the RNA molecules of these two viruses differ in number and function? In your response, consider the role of both in the formation of mRNA.

4. Compare and contrast the reproductive life cycle of WNV (Figure 3.4) to that of HIV (see Figure 19.8 in your text).

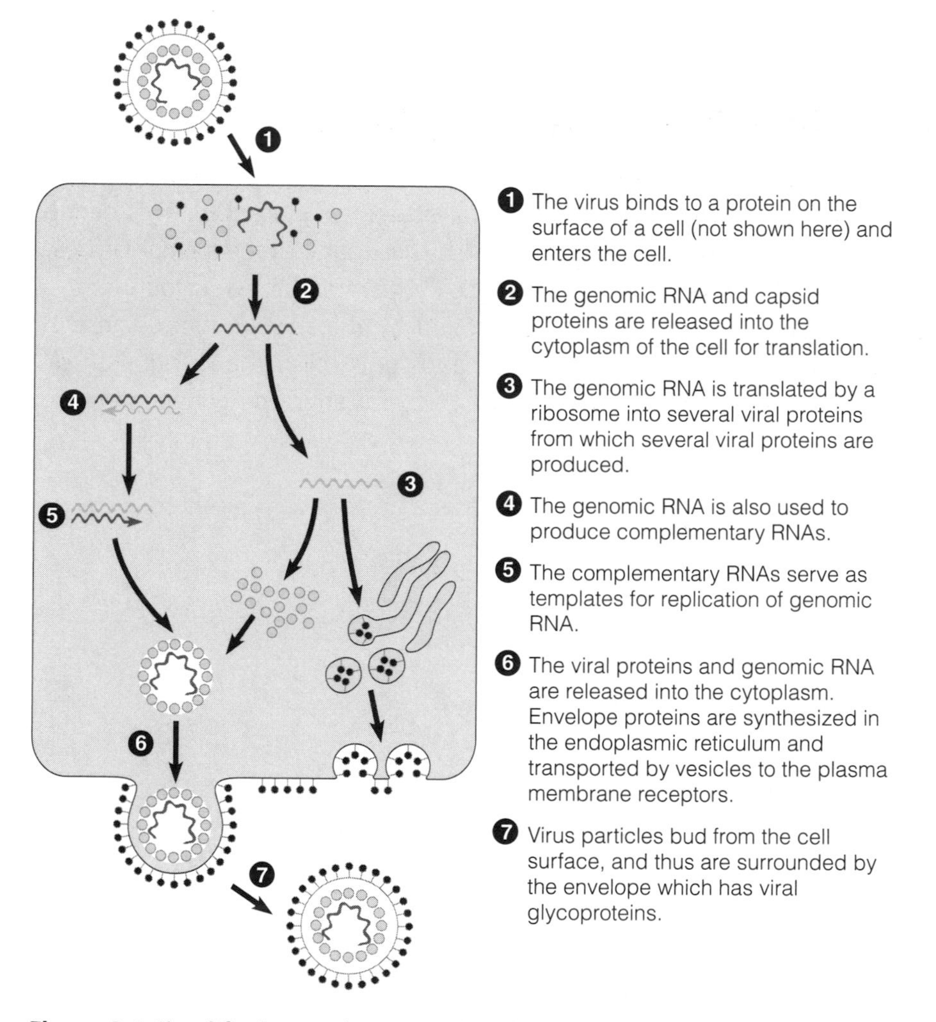

Figure 3.4 Simplified reproductive cycle of the West Nile virus.

5. Many viruses, including WNV, cold viruses, and flu viruses, reproduce in the host for a short time before being destroyed by the host's immune system. This production of new virus particles occurs during a period in which virus particles are present in the blood (viremia). If Russell, the blood donor in the case, had been infected with WNV, he could safely make future donations once the viremia had passed and his blood no longer contained virus particles. In contrast, a person infected with HIV can never give blood. Examine the life cycle of HIV and suggest a reason for this. (*Note:* An immune system response is usually initiated by recognition of "nonself" molecules on the surface of infected cells.)

IV. Testing Blood Donations for WNV

To prevent human blood-to-blood transmission of WNV, all blood donations since June 2003 have been tested for the presence of WNV particles. The test used is called reverse transcription–polymerase chain reaction (RT-PCR).

A PCR cannot be run without DNA. Since WNV does not contain DNA, its RNA must be isolated and reverse transcribed (RT) to form complementary DNA (cDNA). (See Figure 20.8 for more information on PCR and Concept 20.4 for more information on RT-PCR.) When donor blood is tested for WNV, RNA is extracted from the blood sample. Individuals who are in the viremic phase of WNV will have West Nile virus RNA present in their blood, as well as other types of RNA, including their own. The mixed sample of "unknown" RNAs is reverse transcribed to create a mixed sample of cDNAs.

PCR utilizes polymerase enzymes and specific DNA "primers" to amplify (make many copies of) a targeted DNA sequence. Primers are short, single-stranded DNA molecules that match up to the two ends of the targeted DNA sequence and are necessary for the initiation of DNA synthesis. The DNA that matches up to the primers is then repeatedly duplicated in cycles of PCR, until it reaches detectable levels.

Primers specific for WNV cDNA are used in the PCR test referred to in this case. If WNV is present in the blood sample, then the cDNA will be amplified successfully. The primers ensure that a fragment will be amplified from this cDNA only. (For more information, see Khanna et al., cited in the references at the end of this investigative case.)

1. Why are primers needed for initiation of DNA synthesis using PCR? How do PCR primers differ from the primers in cells? (Hint: See Figure 16.16.)

2. The following cDNA sequences (A–D) were obtained by reverse transcription of RNA samples from donated blood. One of the WNV primers used in RT-PCR has the following sequence.

3′ GGCTGCTGGCAACTT 5′

Circle the cDNA sequence below that would be targeted by this WNV primer.

a. 5′ GGCTGCTGGCAACTT 3′

b. 5′ CCGACGACCGTTGAA 3′

c. 5′ TATAACCGTCCAAGTT 3′

d. 5′ CCGGCCTAGCATAGAA 3′

3. Explain how primers control which cDNA is being amplified.

4. The day after Russell's blood sample was tested for WNV, he was told that the results were positive. What organisms were likely involved in Russell's infection with WNV? Is it likely he will pass on the disease?

Additional Investigations

V. Tracking West Nile Virus

A. Origin of the West Nile Virus in the United States. WNV was first isolated in Uganda in 1937 and has since spread throughout Africa and other parts of the world. As an emerging disease, WNV continues to generate both public and scientific interest. Researchers are exploring questions about its origin, evolution, transmission by multiple vectors and host tissues, replication in multiple hosts, detection, and vaccine potential. Central to these investigations are the use of molecular data, including nucleic acid sequences, and the use of bioinformatics (the application of computer science and mathematics to genetic and other biological information).

When WNV was first detected in New York City in 1999, researchers wanted to know where it came from and how it arrived. To propose an answer to these questions, using methods similar to those used in the analysis in Investigation II, you can look at similarities between a New York strain (NY99) of WNV isolated from a Bronx Zoo flamingo in 1999 and strains of WNV isolated from different parts of the world. (*Note:* In Table 3.3, only a portion of the genomes were compared—specifically, a portion of the envelope protein gene. A software program called CLUSTALW was used to align the nucleic acid sequences found in these strains, and then a second program called BOXSHADE was used to display the sequences from the most similar to the least similar compared to the NY99 strain. The Case Book website provides links to instructions for using these programs.)

Table 3.3 BOXSHADE Plot of Aligned WNV *E* Gene Sequences from Various Strains

The BOXSHADE program automatically generates several colors to indicate properties of nucleic acids. To learn more, go to the Biology WorkBench website (see References).

NY99	CCAACTACTGTGGAGTCGCACGGAAACTACTCCACACAGGTTGGAGCCACTCAGGCAGGGAGATT
ISRAEL98	CCAACTACTGTGGAGTCGCACGGAAACTACTCCACACAGGTTGGAGCCACTCAGGCAGGGAGATT
MOROCCO 96	CCAACCACTGTTGAGTCTCATGGTAACTACTCCACACAGATTGGGGCCACTCAGGCAGGGAGATT
ITALY98	CCAACCACTGTGGAGTCGCATGGAAACTACTCCACACAGATTGGGGCCACTCAGGCAGGGAGATT
SAFRICA99	CCAACCACTGTGGATTCGCATGGTAACTACCCCACACAGATTGGGGCCACTCAGGCAGGGAGATT
ROMANIA96	CCAACCACTGTGGAGTCGCATGGAAACTACTTCACACAGATTGGGGCCACTCAGGCAGGGAGATT
TAJIKISTAN99	CCAACCACTGTGGAGTCGCATGGAAACTATTTCACACAGATTGGGGCCACTCAGGCAGGGAGATT
MADAGASCAR88	CCGACGACTGTTGAATCTCATGGCAATTATTCAACACAGGTTGGGGCCACCCAGGCTGGAAGATT

1. Scientists at the Centers for Disease Control and Prevention (CDC) concluded that NY99 most likely was transported to New York from Israel. Does the information in Table 3.3 support this conclusion? How many differences in sequence are there between the two samples? What other conclusion could you draw from comparing the NY99 and ISRAEL98 strains?

2. Which strain is the most dissimilar to NY99? How many differences did you find between this strain and NY99? Do you find this result surprising? Explain.

3. How do you think WNV arrived in New York City? Consider what you've learned previously about transmission of this disease.

B. Spread of WNV in the United States. Since 1999, WNV has been carefully monitored. The CDC maintains resources including regional data and maps to track the spread of WNV in the United States. For example, the map in Figure 3.5 reflects both vector (mosquito) and host (birds, horses, humans, and so on) data collected by the CDC. Human cases reported in any state from 1999 through 2002 are distinguished by cross-hatching.

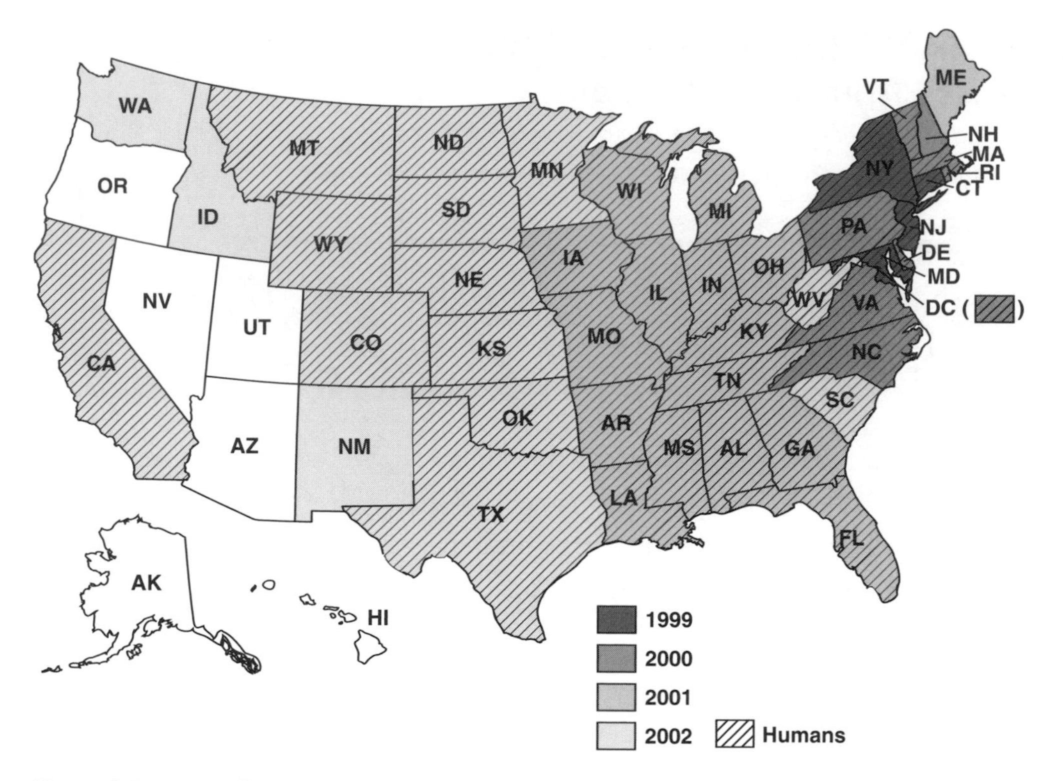

Figure 3.5 West Nile virus in the United States (1999–2002). (Source: CDC)

1. Construct a line graph that shows the number of states reporting the presence of WNV from 1999 through 2002.

2. Is proximity to known outbreaks of WNV a factor in its spread? Looking at the map in Figure 3.5, describe geographic factors that seem to influence the spread of WNV. Explain.

3. Examine the map in Figure 3.6 and compare it to that shown in Figure 3.5. In three to four sentences, describe the extent of spread in 2006.

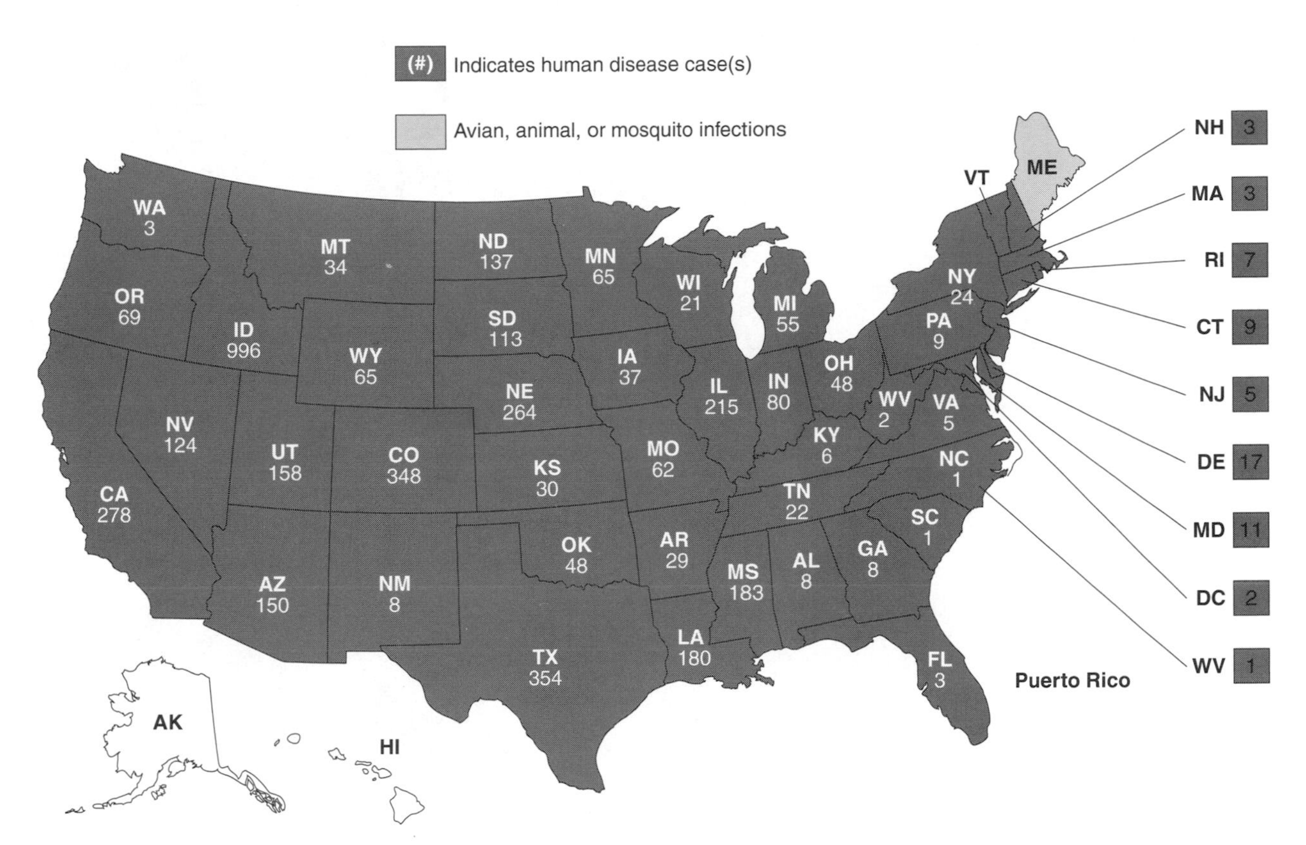

Figure 3.6 West Nile virus activity in the United States (2006), CDC: http://www.cdc.gov/ncidod/dvbid/westnile/Mapsactivity/surv&control06Maps.htm

VI. Open-Ended Investigations

You may wish to visit the West Nile Virus Problem Space to use tools, methods, and data to explore the global spread and evolution of WNV.

The West Nile Virus Workbench Lab (Kiser 2004) provides instruction on using the data and bioinformatics tools.

Additional Potential Investigations are listed in the WNV Problem Space at http://bioquest.org/bedrock/problem_spaces/wnv/curr_resources.php.

References

Berthet, F.-X., H. G. Zeller, M.-T. Drouet, J. Rauzier, J.-P. Digoutte, and V. Deubel. Extensive nucleotide changes and deletions within the envelope glycoprotein gene of Euro-African West Nile viruses. *Journal of General Virology,* 78(9):2293–297, 1997.

Note: Table 3.2 presents an alignment of published DNA sequences of WNV, edited for length. We obtained these sequences from GenBank using identifiers provided by Berthet et al. (see reference above). The sequence identifiers are: EGY-HEg101/51, FRA-PaH651/65, SEN-AnD27875/79, SEN-ArD78016/90, UGA-MP22/?, and MAD-ArMg956/86. We then used the nucleic acid alignment tool CLUSTALW on these sequences. The Biology Workbench was the interface that provided the nucleic acid tools and access to the San Diego Supercomputer. It may be freely accessed at http://workbench.sdsc.edu.

Khanna, M., K. J. Henrickson, K. Harrington, C. R. Waters, J. Meece, K. Reed, and S. Shukla. "Multiplex PCR–EHA Compared to 'Real Time' Taqman for the Surveillance and Diagnosis of West Nile Virus." Prodesse Inc., Waukesha, WI, Medical College of Wisconsin, Milwaukee, and Marshfield Clinic Research Foundation, Marshfield, Wis. Presented at the 11th International Conference on Infectious Diseases, March 2004, in Cancun, Mexico. http://www.prodesse.com/resources/ICID_2004_WNV.pdf

Kiser, Stacey. West Nile Virus Workbench Lab. 2004. http://bioquest.org/bedrock/problem_spaces/wnv/curr_resources.php (accessed July 2, 2007).

CHAPTER 4

Tree Thinking

Tree Thinking

Teruko and her friend Sean were jogging around a track after work. "So, Teruko, how was your brother's wedding in Japan?" Sean asked.

"It was amazing! I couldn't believe all the ceremony," Teruko said. "The reception had some unusual food prepared by special chefs. My favorite was the kujira."

"What's kujira?" Sean asked.

"It's whale meat," Teruko replied. When Sean made a face, Teruko continued. "I know it sounds awful, but it tasted so good. In fact, my dad even tried to bring some canned whale meat back, but customs agents took it at the airport."

"Why?" Sean asked. "Isn't whaling legal in Japan?"

"Sort of. This happened at U.S. Customs. It turns out that you can't bring in any product that is made from marine mammals because so many species are endangered," Teruko explained.

"Oh, that makes sense," Sean said. "So, did your dad get into trouble at the airport?"

"No, they just took the cans away," Teruko sighed.

"I've heard there's a huge black market for whale meat in Asia, and some people pay the equivalent of $400 a pound," Sean added.

"Yes, and they even have biotech tests now to tell if the meat is really from whales."

"How can they tell what type of meat is in the can?" Sean asked.

"Well, they extract DNA from the sample and compare its sequence to known DNA whale sequences. In fact, forensic DNA testing of 'whale meat' from Asian markets has turned up dolphin, shark, and even horse meat."

"Hm. Maybe they should run the same tests on the meat we had at lunch yesterday," Sean laughed. Now it was Teruko's turn to make a face.

Figure 4.1 Minke whales are found in the Atlantic, Pacific, and Antarctic oceans.

CASE ANALYSIS

1. **Recognize potential issues and major topics in the case.** What is this case about? Underline terms or phrases that seem to be important to understanding this case. Then list **three or four** biology-related topics or issues in the case.

2. **What specific questions do you have about these topics?** By yourself, or better yet, in a group, list what you already know about this case in the "What Do I Know?" column. List questions you would like to learn more about in the "What Do I Need to Know?" column.

What Do I Know?	What Do I Need to Know?

3. Put a check mark by **one to three** questions or issues from the "What Do I Need to Know?" list that you think are most important to explore.

4. **What kinds of references or resources would help you answer or explore these questions?** Identify two different resources and explain what information each resource is likely to give that will help you answer the question(s). Choose specific resources.

Core Investigations

I. Critical Reading

To complete this investigation, you should have already read Chapter 26: Phylogeny and Systematics.

A. Morphological Observations: Identifying Characters in the Dendrogrammaceae. In this exercise, you will observe and record morphological characters and investigate the construction of a cladogram based on five characters.

Begin by reviewing Figure 4.2 and Table 4.1. Then record the presence or absence of the five characters that are examined in Table 4.1. A "0" indicates that a taxon does not exhibit the particular character. Enter a "1" if the taxon does exhibit a particular character. For example, taxa B, C, and D have narrow leaves, so a "1" has been entered in the table.

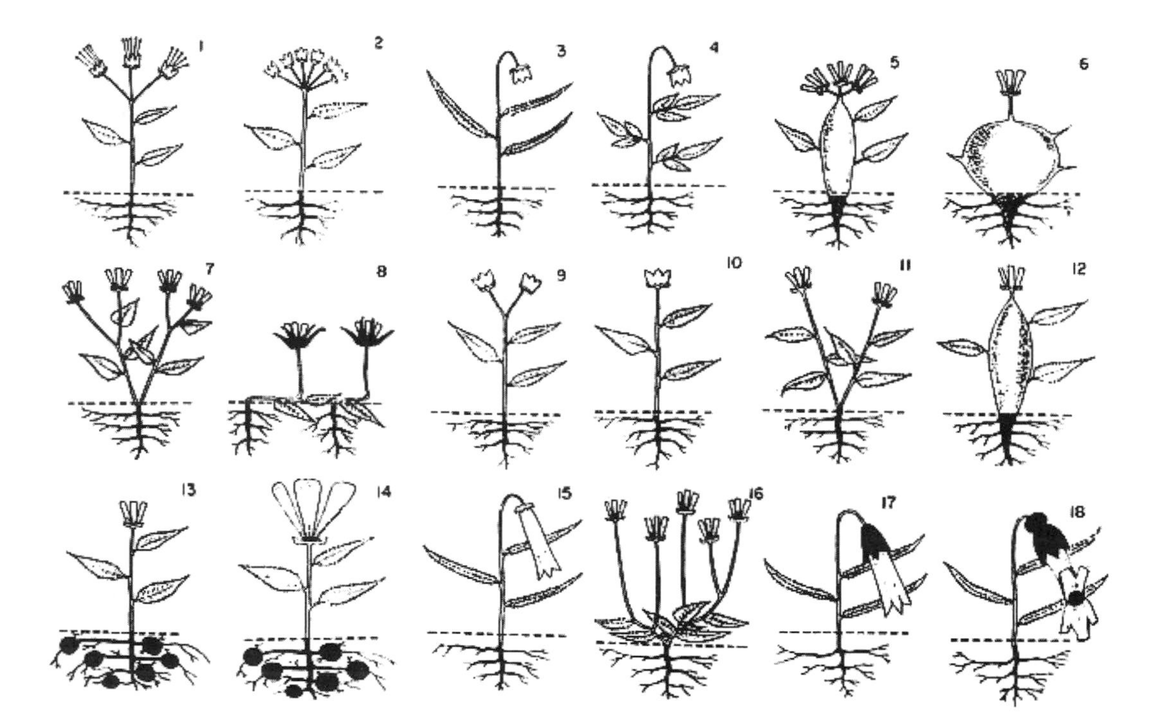

Figure 4.2 The Dendrogrammaceae, an imaginary family of flowering plants (Wagner, W. H., Jr., 2001).

Table 4.1 Observation and Identification of Morphological Characters

(Refer to your text for definitions of unknown terms.)

Selected Taxa	1 Leaves Narrow	2 United Petals, Not Separate	3 Tubelike Petals	4 Elongated Sepals	5 Flower with Bilateral, Not Radial, Symmetry
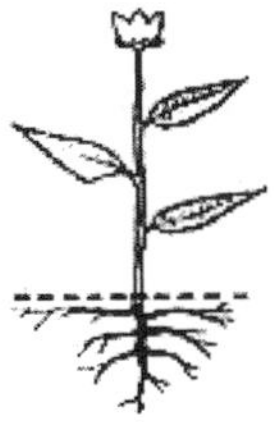 A	0				
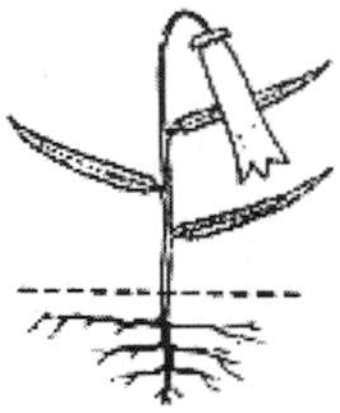 B	1				
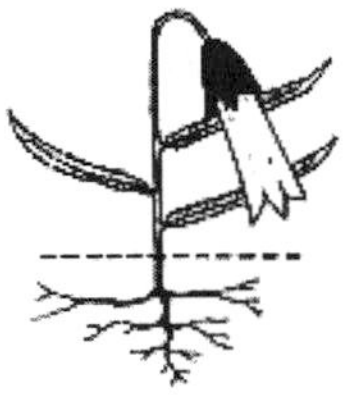 C	1				
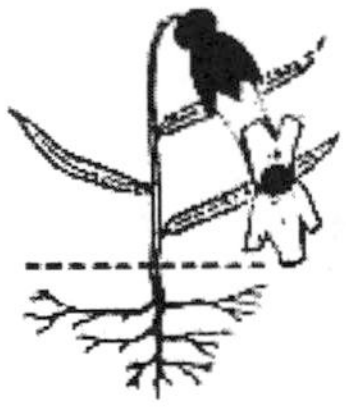 D	1				
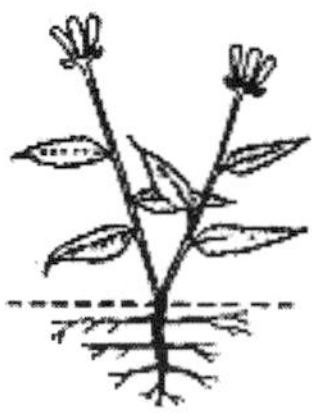 E	0				

1. Which trait in Table 4.1 is shared by at least four taxa? Which taxa are they?

2. All five characters are present in which taxon?

3. Are any of these characters shared by all five taxa?

4. One methodology that has proven useful in developing cladograms is to include a taxon that is less related to the other taxa. This "outgroup" is assumed to have ancestral forms of the characters found in these taxa. Characters that are not shared with this "outgroup" taxon are considered derived. Which taxon is the outgroup in this plant family example?

B. Examining a Cladogram. By constructing a cladogram using the morphological characters recorded in Table 4.1, you can infer relatedness among the taxa. Cladograms can be helpful depictions of patterns in levels of relatedness for shared characters among taxa. Taxa are sorted by presence or absence of characters. However, branch distances in a cladogram do not imply chronology. (Phylogenetic trees present hypotheses about the evolution of taxa and imply chronology in diverging branch points.)

Using the characters in Table 4.1, the following cladogram indicates the relationships among the five taxa selected from the Dendrogrammaceae. Notice that there are branches in Figure 4.3 associated with each taxon. Each node is called a clade. Branches C and D are nested within the larger clade that includes B.

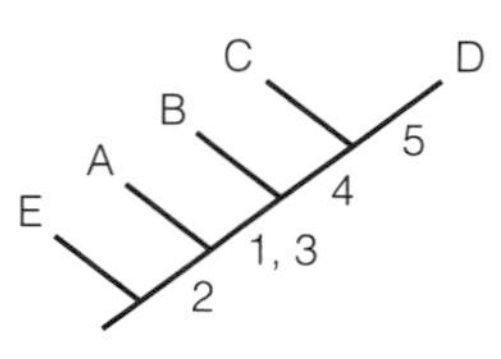

Figure 4.3 The characters are placed below the diagonal. The taxa A, B, C, D, and E are positioned in branches along the diagonal to indicate which characters they share.

1. Which character in Figure 4.3 is exclusively shared by taxa C and D?

2. Consider character 1 in the cladogram. Which taxa lack this character?

3. Can you estimate how much time passed between the development of character 2 and character 4? Explain.

4. Compare the table of characters to the cladogram. Do cladograms contain the same information as the table? Are cladograms more useful? Explain.

C. Identifying Primitive Versus Derived Characters. Simple observation helps differentiate morphological characteristics of these plant taxa, but what characters can help us understand the evolutionary relationships between taxa? To be useful for cladistics, characters must be homologous (reflect shared ancestry). Among the homologous characters, we need to identify the following:

- Shared primitive characters: homologous traits that are common to larger taxonomic groups; for example, flowers are found in all angiosperms.
- Shared derived characters: homologous traits that are limited to particular taxa; for example, flowers with united petals are found in only some angiosperms.

Depending on the taxa included in a cladogram, the same character could be considered primitive in one cladogram and derived in another. For example, consider milk production of mammals. When the taxa are all mammals, then milk production is a primitive character. However, when the taxa include reptiles, birds, and mammals, then milk production is a derived character.

1. Look at the complete group of taxa in Figure 4.2. Select a taxon with a morphological character (other than those you used in Table 4.1) that you think is a shared derived character. What makes it likely to be derived?

D. Revising a Cladogram. Systematists use existing data or gather new data to consider carefully which characters to use in constructing a cladogram. Determination of the branch points is based on these characters. As new data are discovered, a cladogram may be reconstructed to incorporate the new information.

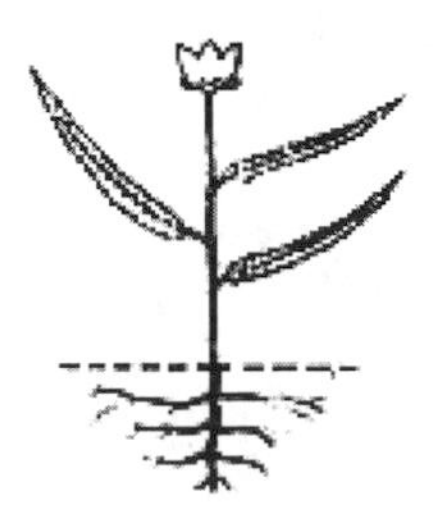

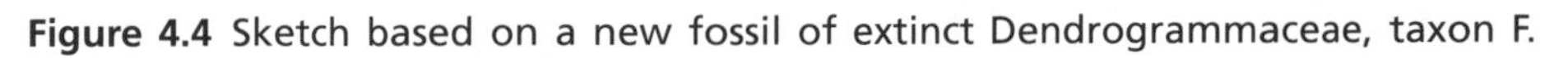

Figure 4.4 Sketch based on a new fossil of extinct Dendrogrammaceae, taxon F.

1. See Figure 4.4 to observe characteristics of a new fossil, taxon F. Enter your observations of taxon F in Table 4.2. *Note:* These are the same five characters used in Table 4.1.

Table 4.2 Character Table for Taxon F

Selected Taxa	1 Leaves Narrow	2 United Petals, Not Separate	3 Tubelike Petals	4 Elongated Sepals	5 Flower with Bilateral, Not Radial, Symmetry
F					

2. Use the character information from taxon F to redraw the cladogram (Figure 4.3).

II. "Whale Meat Forensics"

A. Using Biotechnology and Systematics. In this investigation you will be working with DNA sequence analysis. You should be familiar with the DNA analysis methods covered in Chapter 20.

People in many regions of the world rely on animals harvested from the sea as a significant source of protein. Several cultures enjoy traditional dishes made from whale meat (Figure 4.5). With declining populations of cetaceans (the mammalian order to which whales and their close relatives belong), however, this practice has come under international scrutiny.

Figure 4.5 Several dishes made from different cuts of whale meat.

From 1993 to 1999, researchers from New Zealand (Baker et al., 2000) analyzed genetic variations in a defined region of cetacean DNA. One of their goals was to identify food products containing meat from protected or endangered cetacean species or noncetacean sources. Investigators began by taking tissue samples from beached and harvested cetaceans. Species identifications of the animals were done onsite by experts in cetacean systematics using multiple morphological characters. DNA from the identified cetacean tissue samples was extracted and the targeted DNA was sequenced. The researchers then sequenced the DNA of samples from whale products ("unknowns") sold at retail markets in Japan and the Republic of (South) Korea. By comparing the genetic sequences of the unknowns with the known sequences, the researchers could infer the similarity and species identification of the meat in the "whale" products.

Analysis of 655 products revealed meat from baleen whales (eight species or subspecies), sperm whales, pygmy sperm whales, beaked whales (two species), porpoises, killer whales, dolphins (numerous species), and sharks, as well as from domestic sheep and horses (Baker et al., 2000).

1. What types of biotechnology techniques were used to investigate products sold as whale meat?

2. The researchers combined systematics with the use of biotechnology to do their forensic work. Define systematics.

3. Why was it important for a systematist to identify the "known" cetacean species from which DNA was extracted and sequenced?

4. Based on your reading in the case, why do you think vendors would substitute horse meat for whale meat in some of the Asian markets?

B. How Is the Analysis of Whale Products Done? This next investigation introduces forensic tools that were used to study whale products, as well as some of the reasoning processes used by systematists to develop hypotheses about relationships.

In our whale meat example, the task of species identification began with isolating DNA from the unknown meat and then looking for a tiny portion of the genomic DNA. A mitochondrial DNA (mtDNA) control region (shaded, Figure 4.6) consisting of only 500 base pairs (bp) was targeted.

Although much of this region is highly conserved (retained with few differences among species), known differences within a hypervariable subsection of the region were used to distinguish among cetacean species.

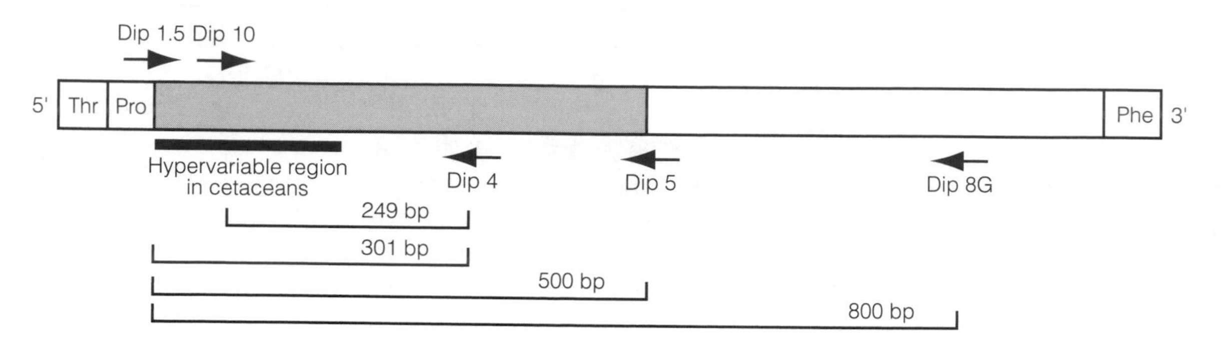

Figure 4.6 A schematic map of the mtDNA control region as well as the binding sites and orientation of the primers used in isolating cetacean DNA. The shaded region represents the portion of the control region covered by most sequences in the reference data sets.

Researchers used PCR to amplify the target mtDNA sequence in the unknown meats. The target mtDNA was then sequenced and aligned with the data set of known cetacean sequences for this segment of DNA. A computer program first compared the target sequence of the sample to known cetacean sequences. Then the program generated a model based on the overall similarities between the target sequence and known cetacean sequences (Figure 4.7).

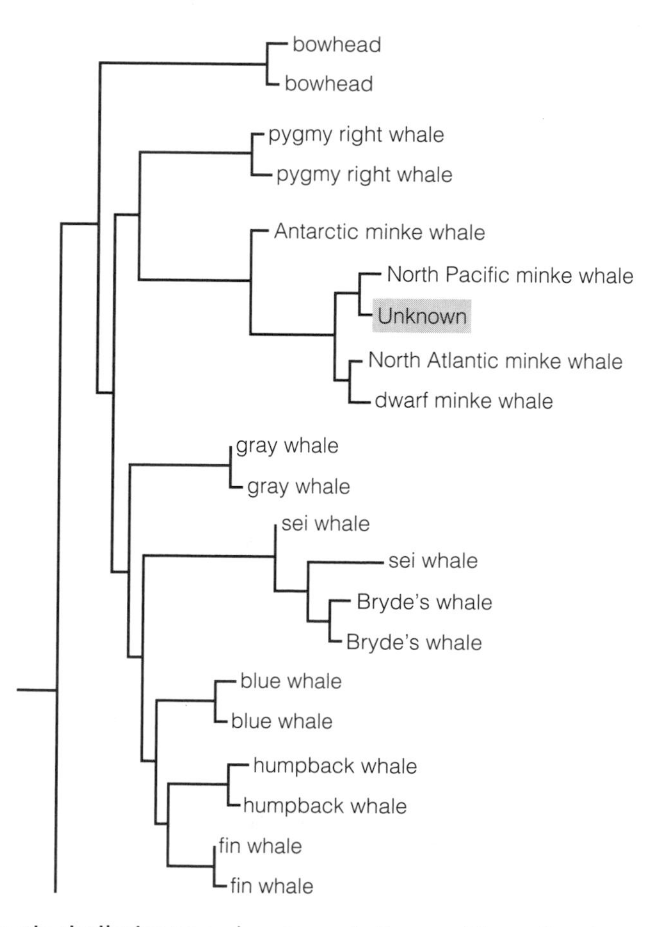

Figure 4.7 Resulting genetic similarity tree showing relative position of unknown samples. Note that the genetic similarity within species is shown by multiple samples in known species, for example, blue whale samples.

1. Which cetacean does the unknown sample most closely match?

The Convention on International Trade in Endangered Species (CITES) works to conserve endangered species by regulating and, where necessary, banning international trade. International trade is banned for those species threatened with extinction. International trade in species that are less endangered or that may become so is permitted when properly regulated. (*Note:* The cetacean identified in your answer to question 1 is found in the second group.)

2. Phylogenetic trees are hypotheses that show a pattern of evolutionary relationships, based on analyses of multiple characters, for multiple taxa. A phylogenetic tree implies a chronological sequence of divergence (branching). Explain why Figure 4.7 is not a phylogenetic tree.

3. Divergence of cetacean species occurred within the last 50 to 65 million years (O'Leary and Geisler, 1999). Explain the difference between the use of a fragment of hypervariable mitochondrial DNA as a molecular marker and the use of a gene such as the one that encodes for hemoglobin. How might you use each of these to compare DNA sequences between cetaceans and other organisms? (*Hint:* See the information in Chapter 26 on molecular clocks.)

4. An attorney defending a whale meat supplier accused of improperly labeling meat would most likely claim that the inferences drawn from the prosecution's evidence were questionable. Provide a potential argument that specifically describes a weakness in the methodology used to infer species identity of the whale meat in this investigation.

5. How might the prosecuting attorney answer this argument? (*Note:* Defend the methodology that was criticized above.)

C. Going Further: Testing Unknowns with "Witness for the Whales." At the website "Witness for the Whales," users can submit unknown mtDNA sequences to be compared against known cetacean sequences. Genetic similarity analyses can be performed. Results are returned in tree and table format, summarizing the genetic distances between the unknown and reference sequences. (Go to the Case Book website for access to and instructions for "Witness for the Whales.")

III. Which Mammals Are Related Most Closely to Whales?

A. The Ungulates. Most scientists agree that whales are members of the ungulates, or hoofed mammals. Some evidence suggests that whales share a common ancestor that gave rise to other living ungulates such as deer, rhinoceroses, horses, camels, pigs, and hippopotamuses. The relationship between whales and other ungulate taxa is still under investigation (O'Leary and Geisler, 1999).

The ungulates are divided into two orders. Horses, zebras, tapirs, and rhinoceroses are odd-toed hoofed mammals, still known as the Perissodactyla. Even-toed ungulates such as deer, sheep, pigs, and cows are known as the Artiodactyla. Until recently, systematists considered Artiodactyla and Cetacea as two distinct mammalian orders. However, several sources of scientific data support the idea that whales are closely related to the members of Artiodactyla. Artiodactyls and cetaceans are now placed in the same order—Cetartiodactyla. The idea of whales sharing an ancestor with these ungulates would seem unlikely if we compared anatomy in living species only.

1. List three characters that you could observe in living whales that white-tailed deer or other even-toed ungulates do not seem to share.

2. Molecular data can also be used to examine relationships between organisms, but the inaccessibility of comparative DNA samples for extinct taxa limits the usefulness of these data. Evidence for shared characters between whales and ungulates based on protein or nucleic acid sequences involves sampling of extant taxa only. However, whale skeletal data incorporate extensive paleontological data from fossils as well as data from extant species. Scientists use fossil data to help reconstruct the hypothetical relationships among whales and other even-toed ungulate taxa. This is a valuable source of data, because like many other taxa, most of the even-toed ungulates that have existed are now extinct.

   Name a character you would expect to find in fossils of early whales that would provide evidence that whales share a common ancestor with other even-toed ungulates.

B. Explore Whale Evolution with the Whippo Problem Space. A good place to begin exploring the relationship between whales and other mammals is the Whippo Problem Space at the BEDROCK bioinformatics education site (see the Case Book website for access information). This Whippo site organizes diverse resources including data and tools to support inquiry.

Consider Figure 4.8. Examine the two trees carefully. Each represents different hypotheses about the evolutionary relationships among the whales and various ungulates. Note that although there are differences in branching patterns (tree shape), both trees include the same outgroup.

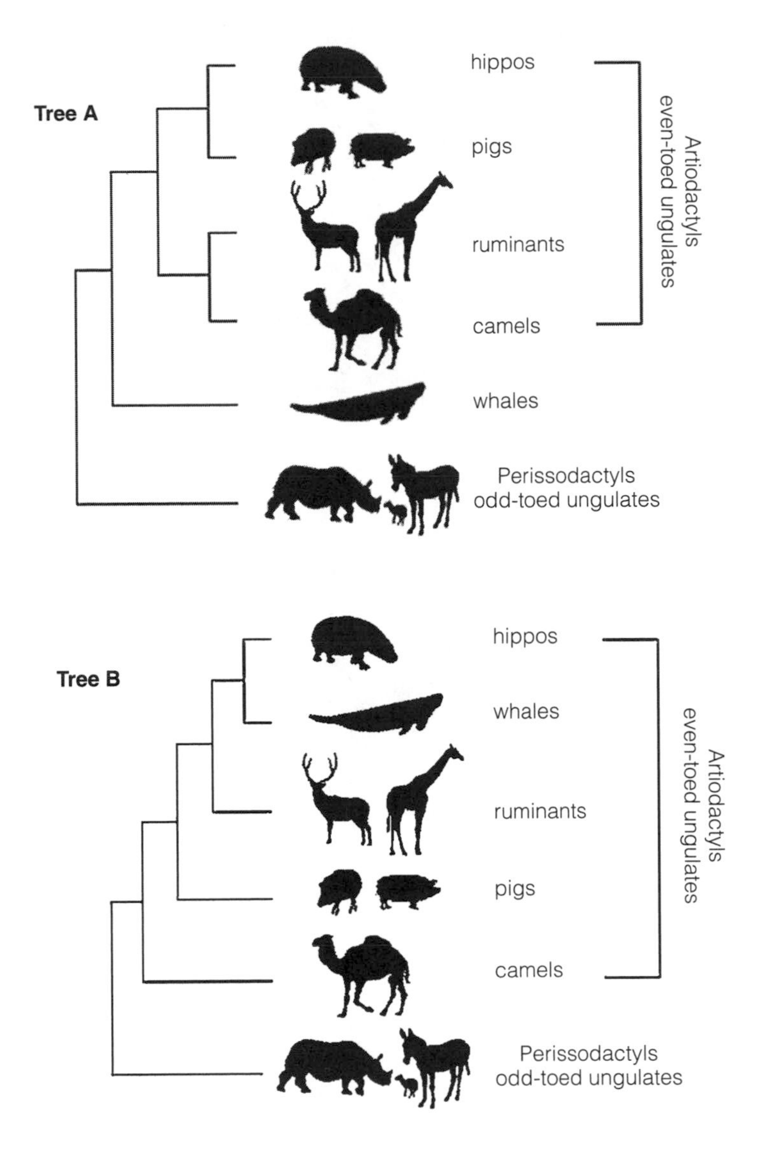

Figure 4.8 Phylogenetic tree representing different hypotheses about the relationships between artiodactyls, the even-toed ungulates. Note the position of whales in each.

1. What is the hypothesized outgroup for both trees?

2. Which tree shows whales and hippos sharing the closest relationship?

3. *Tree thinking* is a term biologists use to describe the process of approaching biological problem solving by considering the role of descent with modification based on phylogenetic evidence. This can result in controversies such as that surrounding the evolution of birds. Did the clade that includes birds diverge from a particular group of dinosaurs? Or did it diverge from a line of reptiles that did not include dinosaurs? These types of questions have fueled much debate and extensive research efforts for decades. Because multiple sources of data exist, both the conclusions reached and the trees drawn by independent researchers may not agree. Not only the selection but also the interpretation of characters

can support different phylogenies. Biologists try to resolve these questions by carefully weighing evidence from multiple sources.

Look closely at the tree shape in Figure 4.9. When you investigate the reasonableness of a tree, you should examine all the hypotheses it contains (Donovan and Hornack, 2004). Each branch point in the tree represents a hypothesis about the relationships among members of the ungulates. The order in which groups diverged is also an explicit hypothesis.

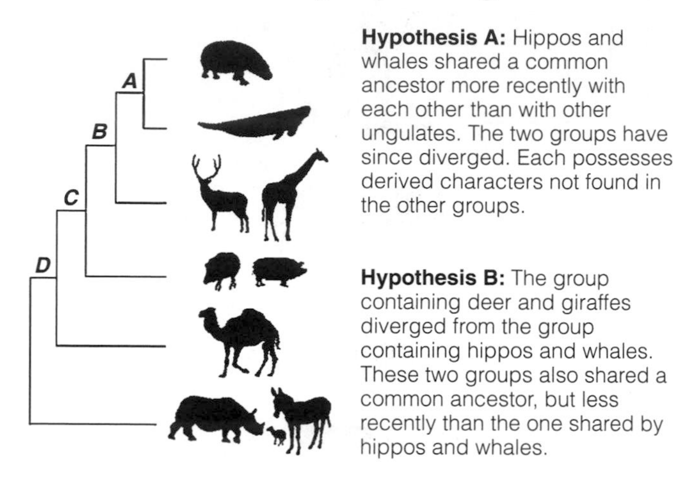

Figure 4.9 Hypotheses in tree B (Figure 4.8) represented by the letters A, B, C, and D.

After you have reviewed the descriptions of hypothesis A and hypothesis B, describe hypothesis D.

4. What does hypothesis D tell us about the relationship between perissodactyls and artiodactyls?

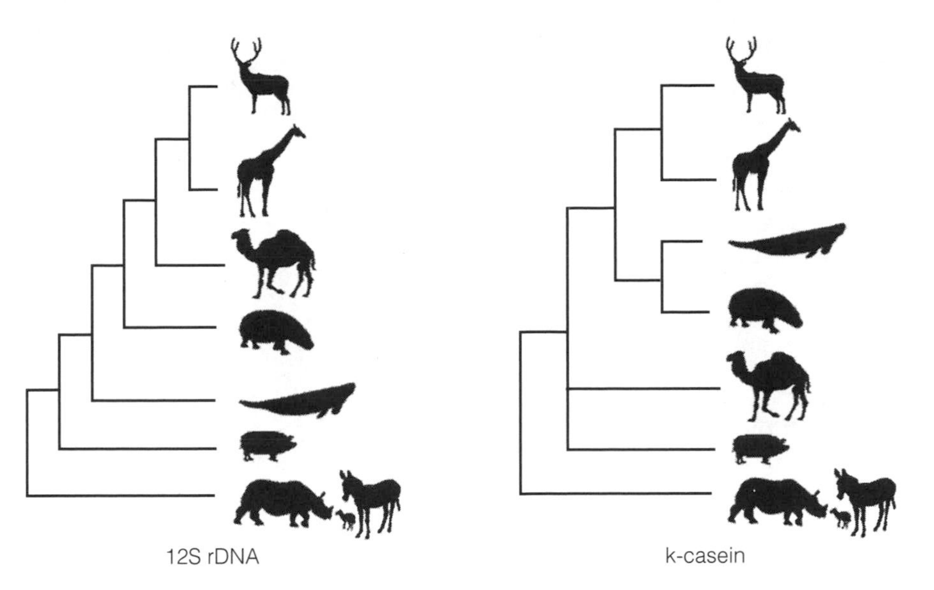

Figure 4.10 Different phylogenetic trees based on select genetic sequences for different molecules. (*Note:* You can read more about tree interpretation on the Whippo site.)

5. Which of the trees based on molecular sequence data shown in Figure 4.10 supports hypothesis A in tree B? Explain.

6. Does either tree in Figure 4.10 support hypothesis B in tree B? Explain.

7. Do you think molecular sequence data are helpful characters to use to infer phylogenetic relationships? What concerns might you have if a tree were based on a single molecule?

Additional Investigation

IV. Position Paper on Whaling

Explore the management of whaling in the modern world. Explain the role of biotechnology and systematics in increasing the global potential for biologically sustainable management of whale populations. Introduce two or more historical, cultural, economic, political, or ethical issues that should be addressed by policies governing whaling practices. Include reliable resources of data on whale population biology.

Reason from information and data to prepare a three- to five-page position paper that specifically addresses your findings about modern whaling policy.

Examples of issues that your paper might address include:

- How are whale populations sampled?
- What ethical issues are raised by the use of biotechnology to police the whale market?
- Should anyone be allowed to whale? Why or why not?
- Which cetaceans should or could be harvested?
- How much might they be harvested?
- Are cetacean sanctuaries feasible?
- Is the U.S. ban on all marine mammal products reasonable?
- What are relevant cultural issues that might need to be considered?

Resources and links to some websites can be found on the Case Book website.

V. Open-Ended Investigations

A. Four new species of additional Dendrogrammaceae have been discovered. How might this change your phylogram? Evolution can include both acquisition and loss of traits, so more than one cladogram may be possible without further information.

B. Consider using Biology Workbench (http://workbench.sdsu.edu) to explore the relationships between cetaceans and taxa other than artiodactyles using genes from the same mitochondrial control region.

References

Baker, C. S., G. M. Lento, F. Cipriano, and S. R. Palumbi. Predicted decline of protected whales based on molecular genetic monitoring of Japanese and Korean markets. *Proceedings of the Royal Society of London B,* 267(1449):1191–199, 2000.

Baker, C. S., M. L. Dalebout, G. M. Lento, and N. Funahashi. Gray whale products sold in commercial markets along the Pacific Coast of Japan. *Marine Mammal Science,* 18:295–300, 2002.

Donovan, S., and D. Hornack. Losing the forest for the trees: Learning to compare trees and assess support for phylogenetic hypotheses. Poster. NABT National Conference, 2004.

O'Leary, M. A., and G. H. Geisler. The position of Cetacea within Mammalia: Phylogenetic analysis of morphological data from extinct and extant taxa. *Systematic Biology,* 48(3):455–490, 1999.

Ross, H. A., G. M. Lento, M. L. Dalebout, M. Goode, G. Ewing, P. McLaren, A. G. Rodrigo, S. Lavery, and C. S. Baker. DNA surveillance: Web-based molecular identification of whales, dolphins and porpoises. *Journal of Heredity,* 94:111–114, 2003.

Wagner, W. H., Jr. Dendrogrammaceae. In E. D. Stanley, "Visual Data Sets," *BioQUEST Library VI.* San Diego: Academic Press, 2001.

Witness for the Whales. http://www.cebl.auckland.ac.nz:9000/page/whales/title (accessed October 10, 2007).

CHAPTER 5

Unveiling the Carboniferous

Memorandum

To: Dericia Williams, Design Team, Wall Works, Inc.
From: Miles Harrington, BioConsulting Associates
Re: Carboniferous Mural
Date: November 19, 2008

Enclosed are comments on the mural design for the new science building at Colter College. The stated goals of the mural project are to: (1) situate well in the local community; (2) relate to the donor's connection to the coal industry; and (3) serve as a resource for teaching science.

The Carboniferous period is an appropriate choice for the mural, as there is a strong scientific connection to the region's history of coal mining and the donor's interests. However, your drawings of a swampy forest of the Carboniferous fall short. A similar conception of the Carboniferous is found in many museums, but it fails to emphasize several significant events of the period. Further, it gives the impression of a single climate—tropical—when in fact there were major glaciations with corresponding drops in sea level and increases in arid zones. Fossils of the period provide evidence of adaptive radiation of previously established terrestrial organisms such as seed plants, tetrapods, and insects.

The climatic changes in the Carboniferous were a significant factor in the evolution of plant communities, which changed in composition during wet and dry periods. During the periods of ice age, the vast swamps diminished as drier upland habitats became available. Instead of tree-sized lycophytes—a common feature of tropical swamps—conifers and seed ferns were dominant species forming the upland forest. Throughout the period, a wide variety of structural adaptations resulted in taller plants.

Although internal fertilization was well established previously, the first amniotes were tetrapods that produced shelled eggs during the Carboniferous. This enabled reproduction as well as embryonic development away from water for both synapsids (reptilelike ancestors to modern mammals) and reptiles. Another major development during the early Carboniferous was flight in insects. A tremendous radiation of flying insects is seen in the fossil record by the end of the Carboniferous.

Although the Carboniferous was a time during which terrestrial habitats were well exploited by an increasing variety of life forms, keep in mind that many of the early adaptations for living on land, such as the insect exoskeleton, tetrapod body shape, and plant vascular tissue arose during earlier geologic periods.

We think that Wall Works, Inc., has a real opportunity to create a mural that richly illuminates the Carboniferous in a way that can serve the regional and educational communities. Let us know if we can be of further assistance.

"Well, this certainly changes things," Dericia thought as she prepared the agenda for the next design meeting. "Maybe we should consider multiple panels and involve the biology department at Colter College." She anticipated lively discussion at the next meeting.

CASE ANALYSIS

1. **Recognize potential issues and major topics in the case.** What is this case about? Underline terms or phrases that seem to be important to understanding this case. Then list **3 or 4** biology-related topics or issues in the case.

2. **What specific questions do you have about these topics?** By yourself, or better yet, in a group, list what you already know about this case in the "What Do I Know?" column. List questions you would like to learn more about in the "What Do I Need to Know?" column.

What Do I Know?	What Do I Need to Know?

3. Put a check mark by **1–3** questions or issues from the "What Do I Need to Know?" list that you think are most important to explore.

4. **What kinds of references or resources would help you answer or explore these questions?** Identify two different resources and explain what information each resource is likely to give that will help you answer the question(s). Choose specific resources.

Core Investigations

I. Critical Reading

You will need to use information from several chapters in Unit Five, as well as the geologic record table in Chapter 25 (Table 25.1), to answer the following questions. You may also want to refer to some of the web links provided for this unit on the Case Book website.

The biology department decided to have students help work on the mural project. The students began by familiarizing themselves with the organisms likely to have lived during the Carboniferous. From looking at images such at Figure 29.16 and reading the caption, the students knew that some insects, as well as lycophytes, horsetails, and tree ferns, were present. Examine Table 25.1 to answer the following questions. You may need to use the index to find information on specific organisms within other chapters of the unit.

1. How long ago was the Carboniferous?

2. List five other terrestrial organisms (not listed elsewhere in this case) that likely lived during the Carboniferous. Include organisms from at least three kingdoms. You may use common names.

3. List four animal phyla that you would expect to find living in the aquatic environments of the Carboniferous, including swamps and oceans.

4. At the first meeting, the artists from Wall Works gave the biology students a list of organisms from their image files. Please evaluate whether the organisms listed in Table 5.1 would be appropriate for the mural of the Carboniferous.

Table 5.1 Appropriate Organisms for Mural?

Type of Organism	Era and Period in Which Organism Is Thought to Have Originated	Present in the Carboniferous? (Yes or No)
Amphibian	Paleozoic, Devonian	Yes
Bird		
Conifer		
Dinosaur		
Grass		
Horsetail		
Human		
Insect		
Jawed fish		
Lycophyte		
Moss		
Mycorrhizae		
Saber-tooth cat		
Spider		
Sponge		
Trilobite		

5. Suggest some reasons for the absence today of dog-size amphibians, tree-size club mosses, and giant dragonflies.

II. The Carboniferous Globe

The Carboniferous, originally named for the rich deposits of coal found in strata of that period in England, is divided into the Lower Carboniferous (or early, often called the Mississippian by researchers in North America) and the Upper Carboniferous (or late, often called the Pennsylvanian by researchers in North America). Limestone deposits laid down on the floors of shallow seas characterize the Lower Carboniferous, whereas coal deposits formed from organisms in swampy forests characterize the Upper Carboniferous. Ongoing glaciation events changed the sea levels during the Carboniferous; therefore, it is not unusual to find some limestone layers interspersed with the coal in Upper Carboniferous deposits.

To get a better idea of what the globe looked like during the Carboniferous, consider Figure 25.13 in the text on the history of continental drift. You could also refer to the Paleomap Project website in the Case Book portion of the Campbell website.

1. Which of the four global landmasses shown in Figure 25.13 corresponds to the Carboniferous?

Additional information is provided by Figure 5.1, which is an evidence-based map of the landmasses during the Upper Carboniferous, with overlay outlines of climate. Several kinds of geologic formations are shown. The small gray circles represent Upper Carboniferous coal beds. The locations of Colter College and its sister college in present-day Scotland are shown.

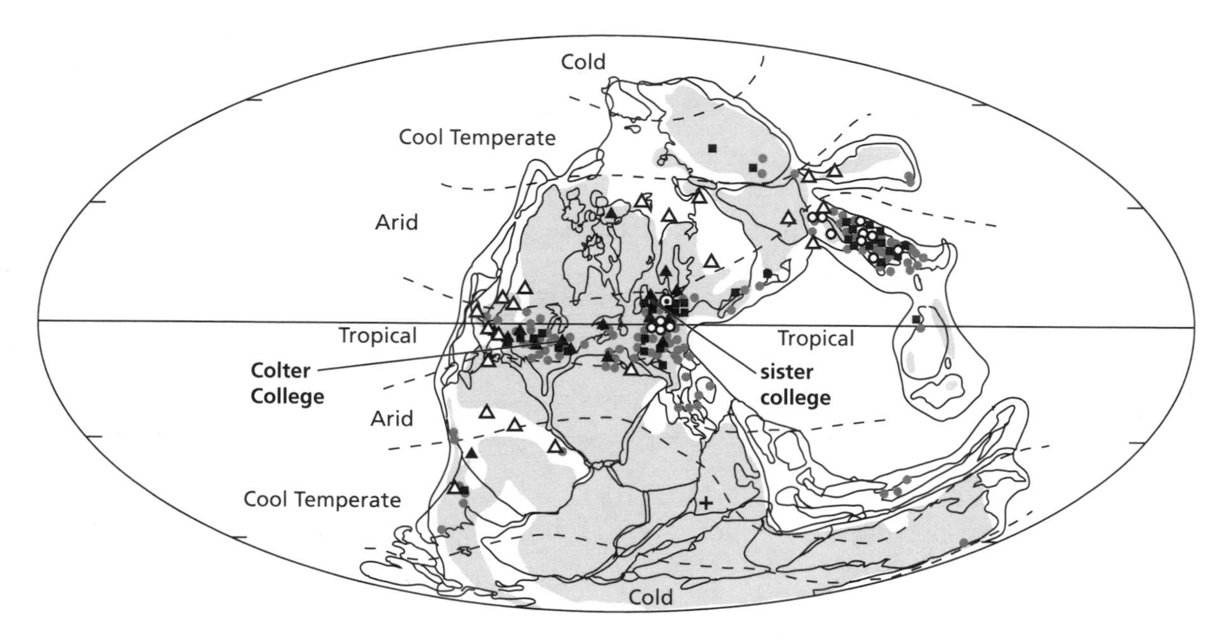

Figure 5.1 The globe during the Upper Carboniferous, showing location of landmasses and climatic conditions.

During the late Paleozoic, the supercontinent Pangaea formed. Not only the position but also the climate of present-day continents were quite different at that time. Not every region had conditions that led to the formation of coal. Use Figure 5.1 to answer the following questions.

2. Consider the map in Figure 5.1 carefully. Are there geographic and/or climatic similarities between Colter College and its sister college in Scotland? Explain.

3. Why don't we find Upper Carboniferous coal deposits in present-day South America, India, Australia, Africa, and Antarctica? What was the climate like in these areas during the Upper Carboniferous?

4. The first recorded use of coal occurred about 3,000 years ago at the Fushun mine in northeastern China. At that time, the people thought of coal as just a remarkable stone that could burn. Based on Figure 5.1, where do you think the land that is now China was located during the Carboniferous? Explain your reasoning.

III. Adaptations in the Carboniferous

Many adaptations contributed to the success of life on land. In the case, Miles Harrington mentioned important adaptations of amniotic eggs and insect wings seen for the first time during the Carboniferous.

1. For the two adaptations listed, attach an image of an organism with this feature that is thought to have existed during the Carboniferous. Write what type of organism it is and explain why the adaptation was significant to the success of the organism.

   Adaptation: Amniotic eggs

   Image Source:

   Type:

   Significance:

   Adaptation: Insect wings

   Image Source:

   Type:

   Significance:

2. During the Carboniferous, forests grew to unprecedented heights. Describe one adaptation in plants that would account for this dramatic change.

IV. Coal Connections

1. Why do we refer to coal as a "fossil fuel"?

In the case, the coal deposits mined near Colter College are from the Upper Carboniferous period, more than 299 million years ago. In highly compressed coal (high-grade coal), the original plant materials and traces of fossils have been obscured. In contrast, coal balls, which are formed when soluble mineral compounds such as silicates and carbonates infiltrate plant tissues, contain fossils with much detail, including microscopic structures (see Figure 5.2).

Figure 5.2 Fossilized plant materials can be observed in remarkably fine detail using an acetate peel of a coal ball. In this image you can see the cross section of a stem.

Analysis of coal ball fossils found in the Colter College area may be helpful because both plant and, occasionally, invertebrate animal fossils may be included. Plant fossils from many Carboniferous coal ball deposits retain permineralized microstructure at the cellular and subcellular levels. Note that it is a matter of chance as to which organisms are fossilized in coal balls. Although it is very difficult to accurately infer population sizes from coal ball analysis, the variety of plant materials seen in a coal ball may provide information about when the coal was formed. For example, if you were to compare coal ball material with modern peat samples you should expect to see significant differences in the plant groups.

2. You observe samples from two coal balls collected in different places. Sample 1 contains roots and stems of tree-size lycophytes and horsetails. Sample 2 contains evidence of sphagnum moss "leaves," yellow cedar needles, and alder pollen. From what time period do you think each sample originated? Explain your answer.

Additional Investigations

V. Calculating Scale Bars and Magnifications

To increase the scientific accuracy and to allow comparisons of the relative sizes of the organisms featured in the mural, you are asked to provide a scale bar or magnification for each of the fossils the artists will be using.

If you work with an image that is its actual size (magnification = 1), drawing a scale bar is fairly straightforward. You would draw a 1-inch line as the scale bar and label the scale value as 1 inch, indicating that 1 inch on the image equals 1 inch of the actual fossil. However, most images are not shown at their actual size because the subject is either too small or too large for the medium. Scale bars, magnification values, and objects of known size allow the viewer to consider the actual size of the subject.

1. You can determine both the image magnification and the size of a fossil specimen if an object of known size is included in the image. In Figure 5.3, the penny has a known diameter of 0.75 inch. However, in this image the penny measures only 0.5 inch across. Now you know that the image is smaller than its actual size. In fact, the magnification is 0.67 times its actual size. (See calculations below.)

   Magnification = diameter of the penny in the image/actual diameter of a penny

   Magnification = 0.50 inch/0.75 inch

   Magnification = 0.67×

   Now estimate the actual length of the leaf fossil using the penny for scale. Use the following calculation.

   Actual leaf fossil length = image length/image magnification

Figure 5.3 Leaf fossil with penny to show scale.

2. Figure 5.4 shows a fossil of footprints made by an anthracosaur, an amphibian from the Upper Carboniferous. We can calculate the magnification of this image because we know that the scale bar (black and white strip) measures 1 inch long with a value of 10.5 cm (each stripe on the bar is 5 mm wide). By doing the calculations below, we can conclude that this image is about one-fourth its actual size.

   Magnification = scale bar length/scale value

   Magnification = 1 inch/10.5 cm

   (*Note:* Convert to same units.)

   Magnification = 2.54 cm/10.5 cm

   Magnification = 0.24×

Figure 5.4 Fossilized anthracosaur footprints.

3. While searching for images, you find a drawing of an early swamp with a dragonfly. The caption says the dragonfly has a wingspan of 2 feet. Use 2 feet as the scale value. The size of the dragonfly wingspan measured from this image is 0.5 inch. Use this value for the length of the scale bar. What is the total magnification of this image?

4. a. Figure 5.5 is an image of a fossilized echinoderm. Estimate the actual diameter of the fossil. (*Note:* Measure from left arm to right arm.)

Figure 5.5 Fossilized invertebrate.

   b. If a penny has an actual diameter of 0.75 inch, how many times would this fossil fit across the diameter of a penny?

5. The wall space for the mural measures 10 feet high by 20 feet wide. You wish to include a 40-m-tall lycophyte and a "giant horsetail" at 15 m tall. To determine which of the magnifications would allow you to fit these organisms in the mural, use the height of the wall as the scale bar length and the height of the tallest organism as your scale value. Which of the following magnifications would allow you to include these organisms?

   (a) 3× (b) 0.5× (c) 0.1× (d) 0.075×

6. You suggest a scale bar that is 1 foot long for the final mural. What is the approximate value of this scale bar in meters?

   (a) 0.4 m (b) 1 m (c) 4 m (d) 10 m

7. Now that you know the dimensions of the mural and the magnification that accommodates the largest trees, what size would a 1-foot-long reptile be in the mural?

8. Artists often use insets (small internal drawings) to change the magnification of a portion of a larger drawing. If you wish to show this reptile and its 1-inch-long egg, what magnification might you choose for an inset? Explain.

VI. Educational Resources to Accompany the Mural

One of the goals of the mural project is to be a resource for science education. After meeting with the artists, the biology students accepted the challenge to develop some resources for explaining the mural to visitors, such as local schoolchildren.

The students proposed a mural-related website and some classroom resources. In addition, they saw the need for an interpretive brochure to be used by visitors of the mural itself. These are good ideas, but they need to be researched, designed, and developed. Develop one of the following resources and submit your project as a paper, poster, slideshow, or another format your instructor suggests.

A. visitor brochure (finished product is two sides of a standard sheet of paper)

B. home page for the website, showing the main categories of links, with one example for each

C. classroom activity, not computer based, indicating grade level, subject matter, materials needed, procedure, and samples of handouts

1. For the resource you selected, answer the following questions.
   - Who is your audience?
   - How do you expect people will use this resource?
   - What are their likely questions?
   - What information can the resource provide to make the mural a richer experience?
   - What are the strengths and weaknesses of this kind of resource?
2. Explain your rationale for the design of the resource. Justify the content decisions as well as the layout decisions. For example, you might argue that it is important to develop a social perspective. This might justify links to a local coal miners' museum included on the website, even though this is not biological information.

VII. Open-Ended Investigations

Use at least three resources to investigate a different post-Cambrian period in order to plan a brochure, mural, time line, or image.

CHAPTER 6

Corn Under Construction

Corn Under Construction

As the local farmers waited for the DeWitt County Extension monthly meeting to begin, they began discussing the surprising news that corn containing a new *Bt* gene not approved for human consumption had been found in a grain elevator in neighboring Macon County.

"I just don't understand it," Emmet said. "Now those growers have to sell all the corn in that elevator as animal feed. How did this happen if no one in that county planted the new *Bt* seed?"

"Well, they may not plant it in Macon County, but we certainly grow all types of *Bt* corn here," John replied.

"I always get some volunteer corn from the year before," Sam offered. "Do you think someone planted new acreage and didn't check for volunteers from the prior owner's crop?"

"I'm pretty sure the new *Bt* seed only became available this year," Emmet responded. "Do you think the seed company might have packaged some of the new *Bt* seed with the wrong label?"

"I know that new *Bt* corn hybrid was planted in at least two fields just north and west of the Macon County border," John declared. "So, what about pollen drift? Remember how windy it was this summer? Why, some of my late-planting corn seedlings in the flood plain were knocked down just about the same time the higher-ground corn was tasseling."

The conversation was interrupted as Roger, the county extension agent, signaled for the meeting to start. "Let's get down to today's business—new alternatives for planting European corn borer refuges—always a popular topic," he announced with a tentative smile.

Figure 6.1 Mature corn plants with tassels and silks present.

CASE ANALYSIS

1. **Recognize potential issues and major topics in the case.** What is this case about? Underline terms or phrases that seem to be important to understanding this case. Then list **3 or 4** biology-related topics or issues in the case.

2. **What specific questions do you have about these topics?** By yourself, or better yet, in a group, list what you already know about this case in the "What Do I Know?" column. List questions you would like to learn more about in the "What Do I Need to Know?" column.

What Do I Know?	What Do I Need to Know?

3. Put a check mark by **1–3** questions or issues from the "What Do I Need to Know?" list that you think are most important to explore.

4. **What kinds of references or resources would help you answer or explore these questions?** Identify two different resources and explain what information each resource is likely to give that will help you answer the question(s). Choose specific resources.

Core Investigations

I. Critical Reading

To complete this investigation, you should have already read Chapter 38: Angiosperm Reproduction and Biotechnology.

A. Reproduction in Corn: Flowers and Pollination. Like the majority of angiosperms, rose family plants have complete flowers. Their floral structure includes sepals, petals, stamens, and carpels (Figure 6.2). If you compare corn flowers to the rose flower, you can observe striking differences. Corn, known globally as *maize,* has unisexual flowers and is monoecious—both male (staminate) and female (carpellate) flowers are found on the same plant. The staminate flowers are located in the tassels produced at the top of the plant. The carpellate flowers are produced in rows on upright ears found lower on the cornstalk.

Figure 6.2 Complete flowers such as this *Rosa* species have both male and female reproductive parts. Maize flowers, however, contain either male or female reproductive parts.

1. Is there any advantage for the corn plant to having its staminate flowers higher than its carpellate flowers? Explain.

2. Each tassel produces 2–5 million pollen grains. One acre of a cornfield may contain 20,000 to 30,000 corn plants, producing up to 68 kg (approximately 150 pounds) of pollen in a single growing season. Each ear has about 1,000 carpellate flowers, although only about 400 seed-containing kernels are produced on the average ear.

The pollen ovule ratio (P/O) in wind-pollinated plants is often greater than 1,000 (1,000 pollen grains:1 ovule). For example, horse chestnut has a P/O of 450,000 and oak has a P/O of 600,000. If 4 million grains of pollen are produced per tassel, what is the ratio of pollen to ovules in an average corn plant bearing one tassel and one ear?

3. Pollen grain size is significant (see Figure 6.3). If a pollen grain is too large, it may not disperse well; however, if it is too small, there will be insufficient resources to produce a pollen tube long enough to reach the ovules. Corn pollen averages 120 μm in diameter, which is much larger than either horse chestnut or oak pollen. Unlike either of these tree pollens, corn has to support the growth of a pollen tube up to 15 cm long.

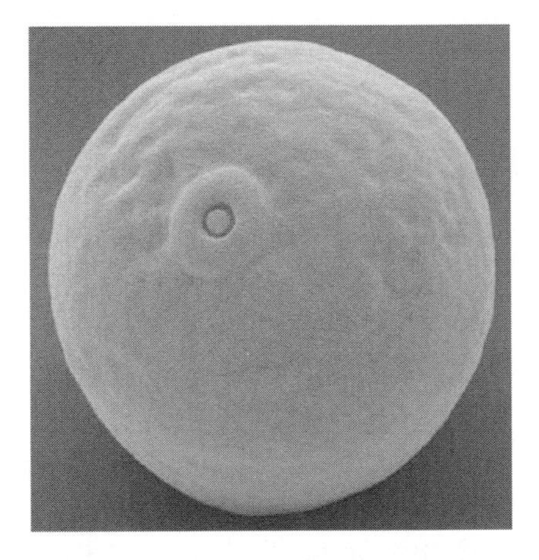

Figure 6.3 A corn pollen grain (~90 μm).

Western ragweed pollen averages 21 μm in diameter and produces a pollen tube length of about 1 cm. Consider the differences in pollen production by corn (4 million grains per plant) and western ragweed (75 million pollen grains per plant). (Western ragweed pollen per stem was calculated from data presented in Wan, S., T. Yuan, S. Bowdish, L. Wallace, S. D. Russell, and Y. Luo, Response of an allergenic species, *Ambrosia psilostachya* [asteraceae], to experimental warming and clipping: Implications for public health. *American Journal of Botany,* 89[11]:1843–846, 2002.) Speculate how the characteristics of each plant's pollen benefit the reproduction of its species.

4. If you didn't know that corn is wind-pollinated, what characteristics of corn flowers could point you toward this conclusion?

5. How do you think a rose is pollinated? Consider your own experience with roses as well as the image provided in Figure 6.2. List two personal observations to support your answer.

6. Roses belong to the clade of flowering plants called eudicots. Corn belongs to the monocot clade. Using your knowledge of eudicot and monocot traits, answer "monocot" or "eudicot" for the following features observed in plants from one of these two clades:
   a. Parallel venation in the leaves

   b. Vascular bundles in the stem arranged in a ring

   c. The seedling produces a single cotyledon

B. Reproduction: Fertilization and the Seed. See Figures 38.3 and 38.8 in the text to help you with this investigation.

1. If you were to slice open a kernel of corn and apply iodine solution to the interior, which part of the kernel do you predict would turn the darkest blue? What is the function of this part of the seed?

2. Do the embryo and endosperm contain genetic information from the female gamete, the male gamete, or both?

3. Do both the embryo and the endosperm have the same number of chromosomes? Explain.

II. Considering *Bt* Corn

A. Misplaced *Bt* Corn. Recall from the case that something strange happened in Macon County. Some of the corn stored in the major grain elevator tested positive for new *Bt* genes. These genes are found in some kinds of genetically modified corn, but this corn was not planted in Macon County according to the cooperative records. (*Note:* Macon County shares some of its north and west borders with DeWitt County.)

1. List the hypotheses posed by the DeWitt County growers as to how new *Bt* corn found its way into the Macon County growers' grain elevators.

2. Consider how the *Bt* genes turned up in the Macon County corn according to John's hypothesis. Within the seeds, would *Bt* genes be found in the embryo, the endosperm, or both? Explain.

3. Farms are spread out all around Macon County. Two members of the growers' cooperative had samples of seed from corn left in their fields that tested for the presence of *Bt* genes. Compare the test results of field A and field B in Figure 6.4.

4. Which field results would tend to support John's explanation of how pollen traveled from DeWitt County to Macon County? Why?

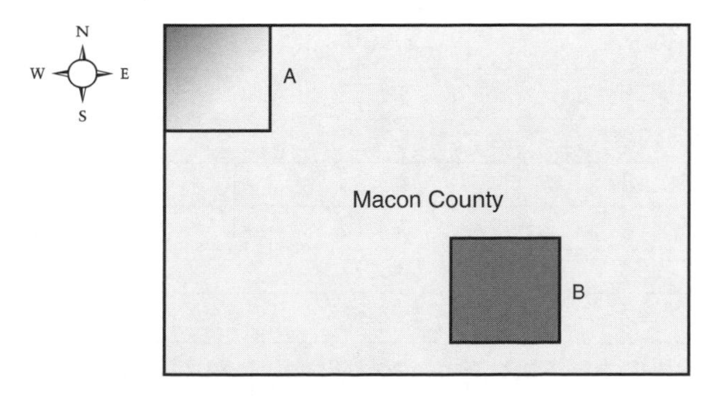

Figure 6.4 Two fields with Bt test results in Macon County. Gray indicates the presence of *Bt* genes in the field. The more corn that tested positive for *Bt* genes, the darker the gray scale.

5. Which field results would tend to support Emmet's explanation? Why?

6. Draw a new rectangle to represent field C with results that would support Sam's hypothesis. Explain the significance of the pattern of *Bt* genes in your sketch.

B. The Economics of *Bt* Corn. The seed for *Bt* corn hybrids costs approximately $14 more per bag than the seed for conventional corn hybrids. This "biotechnology premium" varies from year to year and depends on the type of transgenic seed purchased. The following table describes the potential savings (or losses) of using *Bt* corn under various levels of corn borer populations and corn pricing.

1. Using Table 6.1, estimate the net loss or net gain for a farmer with 2,000 acres in the following scenarios.
   a. Corn prices are low ($1.50) and so are the corn borer populations (about 1 for every 4 corn plants, or 0.25).

Table 6.1 Potential Savings (or Loss) per Acre of *Bt* Corn Versus No Corn Borer Control[1]

Average Number of Borers/Plant[2]	Corn Price per Bushel			
	$1.50	**$2.00**	**$2.50**	**$3.00**
0.00	($4.55)	($4.55)	($4.55)	($4.55)
0.25	($1.99)	($1.13)	($0.27)	$0.58
0.50	$0.58	$2.29	$4.00	$5.71
1.00	$5.71	$9.13	$12.55	$15.97
1.50	$10.84	$15.97	$21.10	$26.23
2.00	$15.97	$22.81	$29.65	$36.49

[1]This table assumes: a yield potential of 144 bushel per acre; *Bt* corn costs $14 extra per bag; a seeding rate of 26,000/acre.

[2]The number of corn borers that would complete development in a non-*Bt* hybrid.

b. Corn prices are high ($2.50) and so are the corn borer population sizes (about 1 for every plant, or 1.00).

2. What other factors might enter into a grower's decision about whether to plant *Bt* corn?

C. Simulations: Hybridization and Genetic Engineering of Crops

1. *Bt* corn is made by replicating the gene for the *Bt* toxin found in the bacterium *Bacillus thuringiensis* and inserting the gene into corn. Techniques described in Chapter 20 enable the plant engineer to identify plants that have incorporated the *Bt* gene. Go to the Case Book website to run a simulation for engineering transgenic tomatoes.
2. Figure 38.16 in the text compares modern corn with its ancestral plant, teosinte. Neolithic farmers selected for traits such as large cobs and kernel size as well as a tough husk encasing the entire cob. Over time, this artificial selection led to the development of modern maize. Go to the Case Book website to use a selective breeding simulation for engineering bigger, better corn.
3. Although both of these strategies are examples of artificial selection of crop plants, describe two differences between these approaches.

D. Alternatives for Controlling European Corn Borers. The European corn borer (ECB) was introduced into the United States in the early 1900s, most likely arriving with imported European plant products. Without predators to keep the population in check, ECBs spread rapidly. In most of the United States this moth produces multiple generations in a single year. Various environmental factors influence the population sizes from year to year. Isolating infested fields and burning the plant material was the only method of control until the mid-1920s (Figure 6.5) when the bacterium *Bacillus thuringiensis* was discovered to have pesticidal properties. By 1930, growers were spraying their crops with a mixture containing live *B. thuringiensis,* which was effective against ECB for as long as the pesticide-producing bacteria survived—several days at most. Although chemical pesticides became widely available after World War II, the majority of growers continued to use the *B. thuringiensis* mixture.

1. Many farmers growing corn using conventional methods still choose to apply *B. thuringiensis* sprays. What are the advantages of this strategy for controlling ECB?

Figure 6.5 In 1919, the only method of controlling corn borers was crop destruction.

2. What are the disadvantages of applying sprays?

III. Investigating Corn Morphology and Growth with a Model of Insect Damage

A. Corn Morphology. Maize is a member of the grass family. As you may recall from Chapter 35, grasses contain meristematic tissue in each node along the length of their shoots as well as in a basal meristem. Although most other plants produce new growth from apical meristems, grass leaves and shoots grow up from the base. Mowing is the equivalent of a "haircut" for grasses, which grow back quickly. Nongrass plants recover from mowing much more slowly, because new apical meristems must form.

1. How does this type of growth in the leaves help grasses survive being eaten by herbivores such as bison?

2. When corn is infested with ECB, several types of damage can occur. Late-season borers may invade the corn ears. Not only does this result in an unsightly appearance and decrease in yield, but the damaged sites are also likely to be colonized by bacteria and fungi. Production of toxic by-products from certain species of fungi such as aflatoxin can result in the entire crop being rejected at the mill.

   Another kind of damage occurs when ECBs tunnel into the solid stalks and create hollow spaces that weaken the plant, which may collapse. Like the damaged ears, damaged stems are also inviting to bacteria and fungi.

   Considering what you have learned about stem structure and function, describe an additional problem that is likely to result from ECB stem tunneling.

To get a better understanding of the structure and growth of grasses like corn, we will study yet a different type of damage from the ECB, called "shot holes."

3. Using Figure 6.6, note the position of each leaf and fill in Table 6.2. Row D is filled in for you.

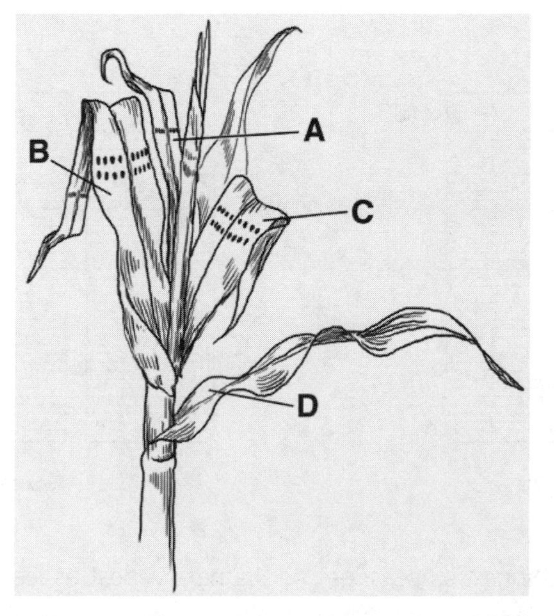

Figure 6.6 A corn plant with early-season damage from European corn borers.

Table 6.2 Damage Done to the Leaves by Corn Borers

Leaf	Damage	Comparative Age
A		
B		
C		
D	None observed	Older than A, B, and C

When the ECB caterpillars first emerge, corn is in an early stage of growth that farmers describe as the *whorl* stage. The new leaves are tightly wrapped around each other at first. Then, as the stem elongates, they separate. The leaves dramatically expand as their cells mature and elongate.

4. Do you think that the damage in Figure 6.6 was caused by several caterpillars feeding on the leaves successively or by one or two feeding at the same time? Explain.

B. Making a Model of Shot Hole Damage to Explore Growth. Models are often useful in exploring complex phenomena by limiting the number of factors involved. Models allow us to simulate interactions, test hypotheses, and ask new questions.

To examine the cause of shot hole damage, you need to first construct a physical model of a young corn plant in whorl stage with three leaves. Then use this model to simulate the feeding activity of corn borers. You need a sheet of ruled notebook paper (8½" × 11"), a ruler, scissors, a writing instrument, and a straight pin.

Part 1: Making the Leaves

Step 1: With the pen or pencil, draw a vertical line at 4 inches and another at 7 inches from the left edge of the paper (Figure 6.7a).

Step 2: Cut the paper lengthwise along the two lines you have drawn so that you have three vertical strips to represent three leaves (Figure 6.7b).

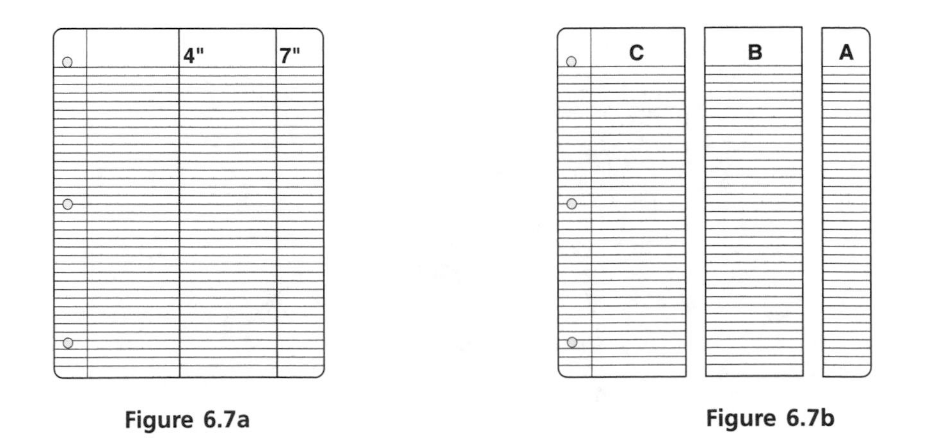

Figure 6.7a

Figure 6.7b

Step 3: Mark the top of the thinnest strip as leaf A, the next widest as leaf B, and the widest as leaf C.

Part 2: Modeling the Age of the Leaves

Step 1: Mark leaf A 10 lines from the bottom (Figure 6.7c). Roll the strip tightly from the bottom until you reach the 10-line mark. Press firmly to fold the rolled paper in place. Leaf A is the youngest leaf. It is the thinnest and the shortest. Much of its maturation and elongation has not occurred.

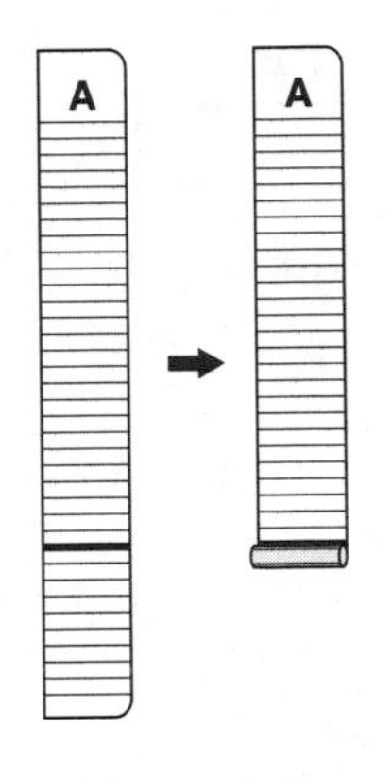

Figure 6.7c

Step 2: Mark leaf B at 5 lines from the bottom. Repeat the roll-and-fold process from step 1. Leaf B is the middle leaf and should be longer and wider than leaf A.

Step 3: Leaf C does not require folding. Leaf C is the oldest leaf and should be the widest and longest. It has completed most of its maturation and elongation.

Part 3: Assembling the Whorl

Step 1: Mark leaf C in the center of the strip 5 lines from the bottom. Label this mark "node 1." Mark leaf B in the center of the strip at 5 lines from the bottom. Label this mark "node 2" (Figure 6.7d).

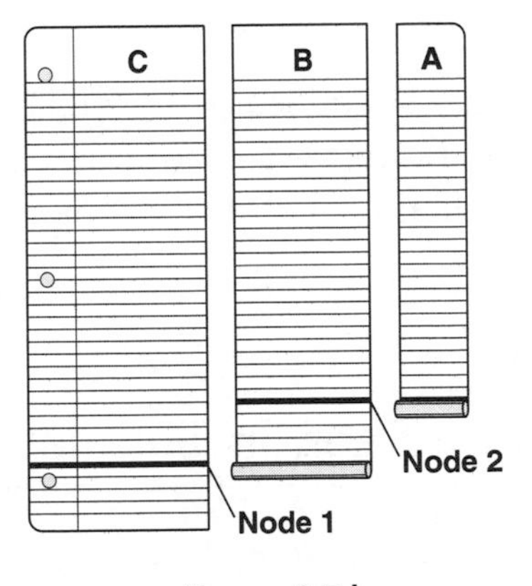

Figure 6.7d

Step 2: Take the folded leaf B and center its base on node 1. Take the folded leaf A and center its base on node 2. The three leaves should now be stacked (Figure 6.7e).

Step 3: Carefully keep the three strips in place (you could use a small piece of tape to secure leaf B and leaf A) as you roll the leaves as one unit from the side around your writing instrument.

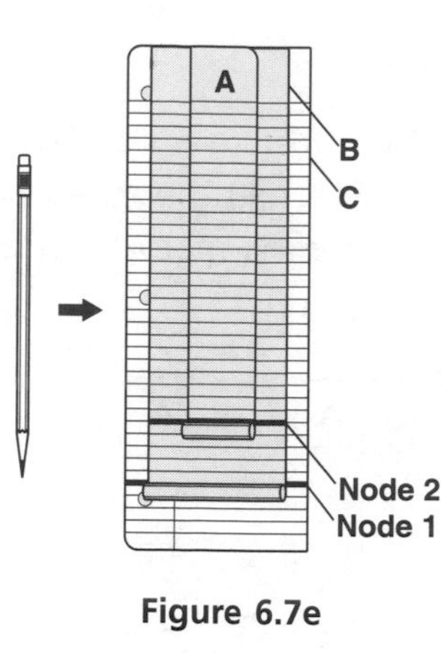

Figure 6.7e

Step 4: Carefully ease out the pencil by pulling it from the center while firmly holding the strips. Then flatten the roll of strips lengthwise for easier handling.

Part 4: Simulating the Corn Borer Damage

Step 1: The resulting model of a corn plant whorl should look tubelike. If you feel along the tube, there will be a noticeable thickening where leaf B begins and another where leaf A begins. These represent nodes on the stem, where the meristems that produce the leaves are located.

Step 2: Locate the flattened area between the node for leaf B and the node for leaf A 2.5 inches or so from the bottom. Take the pin and push it once through the whorl at this location. This simulates the path a borer makes while eating, or "boring," its way straight through the leaves.

Step 3: Locate the flattened area above the node for leaf A 3.5 inches or so from the bottom. Take the pin and push it once through the whorl at this location. This simulates a new path the same borer or a second borer makes while eating its way straight through the leaves.

Part 5: Modeling Growth After the Damage

Step 1: Unroll the model of the damaged whorl and separate leaf A from leaf B and leaf B from leaf C.

Step 2: Unroll the shortened strips of leaf A and leaf B. Carefully replace or reattach the bottom of the unrolled leaves to the nodes that they were attached to before you unrolled them. This repositioning of the leaves models the elongation of both the stem and the leaves during normal growth.

1. Describe the pattern of damage you see on the leaves by filling in Table 6.3.

Table 6.3 Pattern of Damage Seen in Simulated Borer Activity

Leaf	Damage
A	
B	
C	

2. Look again at Figure 6.6 of shot hole damage. Are your model results consistent with this picture?

3. What do you consider to be the limitations of this model of corn growth?

4. What do you consider to be the strengths of this model of corn growth?

5. Do you think making a physical model was helpful in understanding this pattern of damage by European corn borers? Explain.

IV. Refuges for Resistance Management

In the case, we saw that the farmers were going to discuss planting refuges for the European corn borer (ECB). The Environmental Protection Agency requires every grower who plants *Bt* corn to use a refuge so that some *Bt*-susceptible corn borers will survive. When *Bt*-susceptible corn borer moths mature, they are available in adequate numbers to mate with any rare *Bt*-resistant corn borer moth that survives in the *Bt* corn. The eggs produced from these matings are more likely to contain embryos that possess susceptibility genes to *Bt* toxin. In this way, refuges help maintain the gene frequencies for susceptibility in the ECB population and overcome the *Bt*-resistance selection effects found in the *Bt* cornfields. Although resistant populations of ECB are quite likely to develop in the future, the purpose of the refuges is to slow this process. (*Note:* You may wish to use the Hardy-Weinberg equilibrium model to look at the effect of migration on gene frequencies to justify the refuge concept. See the Investigation on the Campbell website in Chapter 23, *How Can Frequency of Alleles Be Calculated?*)

1. In the case, the growers meet to discuss planting refuges in order to reduce the chance that a population of corn borers resistant to *Bt* toxin will become established. Offer two explanations why the development of resistant bacteria is more difficult to control than the development of resistant corn borers. (Hint: See the heading "R Plasmids and Antibiotic Resistance" in Concept 27.2.)

2. All the refuge plans shown in Figure 6.8 provide an interface between *Bt* and non-*Bt* corn. Do you think the block refuge or the split planter refuge would be more likely to facilitate the opportunity for a rare *Bt*-resistant corn borer moth to mate with a moth that isn't *Bt*-resistant? Explain.

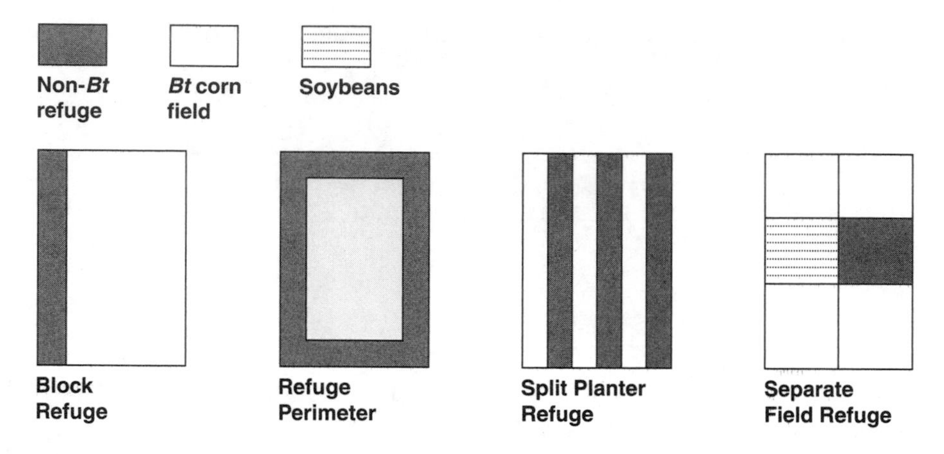

Figure 6.8 Several options for planting refuges. Federal guidelines call for refuges to constitute at least 20% of field space.

Additional Investigation

V. Making Decisions About DNA Technology: Golden Rice

1. Complete the Web/CD Activity: Making Decisions About DNA Technology in Chapter 38. This activity raises a concern that transgenic crops may reduce biodiversity. In your own words, explain how this might occur and why it is significant.

2. Consider your explanation above. Do these risks only apply to transgenic seeds, or is this also true of the hybrid seed that farmers have been using for many years?

3. In developing nations, farmers generally depend on crops that produce both food and seed. Develop arguments both for and against the distribution of *Bt* corn seed to farmers in developing nations. Is a growing reliance on seed companies problematic, or do the benefits outweigh the risks?

   Before you answer this question, consider reading the following two position papers from *Action Bioscience* on the Case Book website:

   "The Ecological Impacts of Agricultural Biotechnology" by Miguel A. Altieri, February 2001.

   "Biotechnology and the Green Revolution," interview with Norman Borlaug, November 2002.

VI. Open-Ended Investigations

A gene from Antarctic fish that allows the fish to avoid freezing has been put into tomatoes. These tomatoes also survive hard frosts.

Find another gene that might be valuable for crop management, enhanced growth, or nutritional quality and tell why. Identify the source of the gene and the target crop.

CHAPTER 7

Galloper's Gut

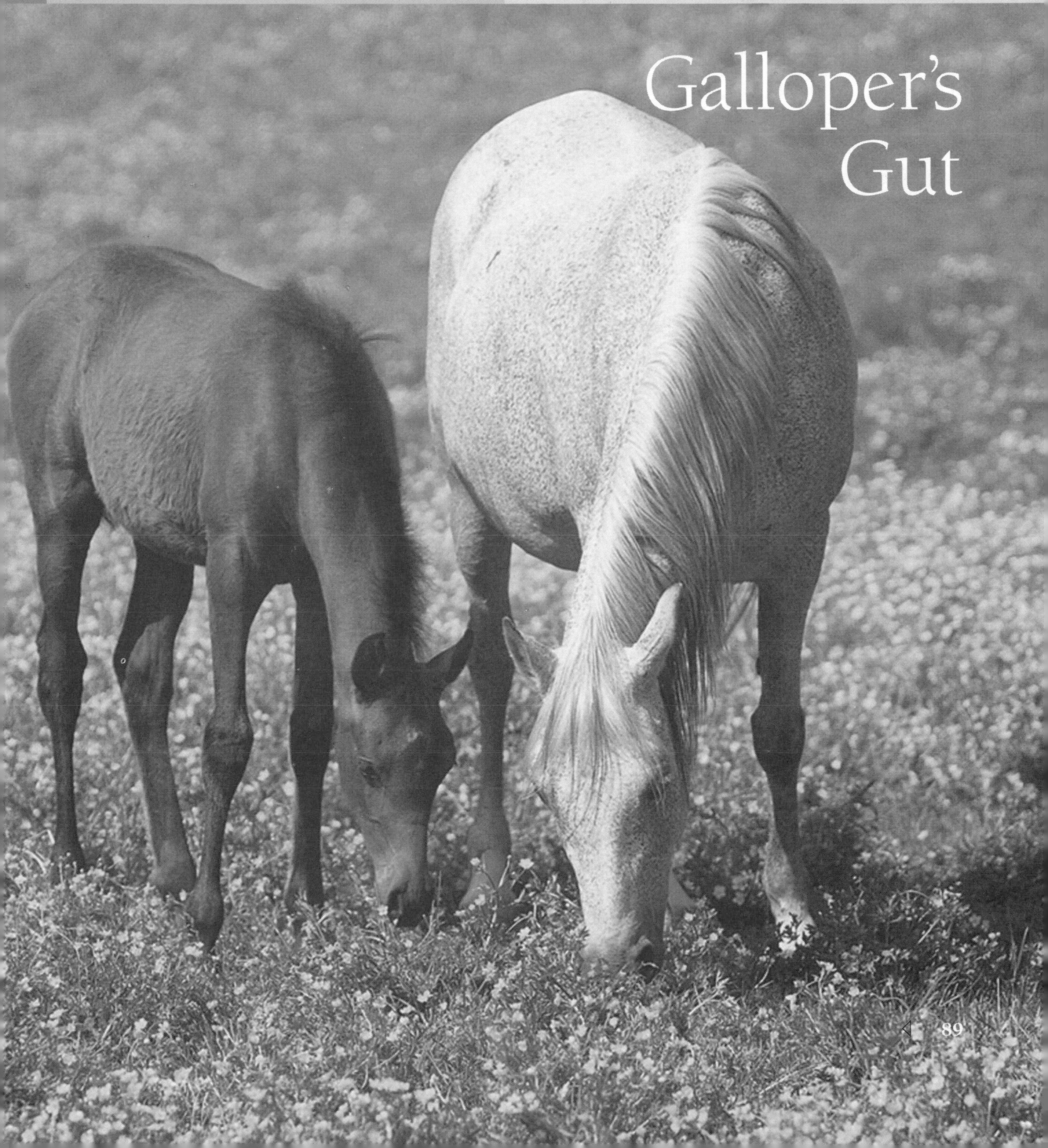

Galloper's Gut

As she walked into the horse barn with Jim and Gina, Leah smiled to herself. Not only did she have a summer job working with horses, but she would get time on horseback as well.

Gina explained, "We have four horses right now. The mare and her foal, a 3-year-old stallion we are training for competition, and this little yearling that is just not growing very well. We have to keep him in the barn all the time to control his feed."

"The grass here is not like the grass we had in Kentucky," Jim declared. "We have to give the horses extra minerals."

"The soil in the pasture *is* a problem," Gina agreed. "But we are replanting with both alfalfa and timothy. Among other things, the local fescue grass turned out to be infected with a fungus. It may be the problem with our little guy, though."

Leah looked around the barn, trying to take it all in. "I didn't get to feed horses before," she said. "What do I do?"

"You'll need to learn to prepare the supplemental feed and give it to each horse twice a day," Jim replied. "We'll show you what to do."

Leah noticed the feeding plan posted on the wall near the bins of feed. "Gosh," she said, "I had no idea it was so complicated. I thought all horses just grazed in the pasture and maybe had oats for a treat. But each of your horses gets a different feed mixture and in different amounts."

"That's right. Each animal has different needs depending on its size and level of activity," Jim said. "For example, we're training Best Boy, so he gets a heavy workout every day. He's an easy keeper, but he does better with some extra oats because of his high activity level."

Gina added, "The mare is lactating now, so she is on a different ratio of hay, grain, and supplements than the others."

"So, do the horses need to go out to the pasture to feed at all?" Leah asked.

"Yes, unless you leave a lot of good-quality hay for them to get at in the barn," Gina explained. "Hay should be green and contain more leaves than stems. The stems should be flexible. Another way I can tell if hay is good is to check if the bale is easy to lift. Heavy bales usually contain water and that might mean mold."

"Horses actually have a very small stomach for their size, so they have to eat almost constantly," Jim explained. "Because of this intense grazing, each horse needs about 2 acres of pastureland to support it. Bad things happen when too many grazing animals are kept on too little land."

"Like that guy down the road with the buffalo?" Leah asked.

"Yes. He has too many animals in that space. His buffalo are definitely too small, and I'd bet they need to be wormed," Gina answered.

"Hey, let's get started with the yearling before we try to solve the buffalo problem," Jim said.

Figure 7.1 A mare and her foal grazing in the pasture.

CASE ANALYSIS

1. **Recognize potential issues and major topics in the case.** What is this case about? Underline terms or phrases that seem to be important to understanding this case. Then list **3 or 4** biology-related topics or issues in the case.
2. **What specific questions do you have about these topics?** By yourself, or better yet, in a group, list what you already know about this case in the "What Do I Know?" column. List questions you would like to learn more about in the "What Do I Need to Know?" column.

What Do I Know?	What Do I Need to Know?

3. Put a check mark by **1–3** questions or issues from the "What Do I Need to Know?" list that you think are most important to explore.
4. **What kinds of references or resources would help you answer or explore these questions?** Identify two different resources and explain what information each resource is likely to give that will help you answer the question(s). Choose specific resources.

Core Investigations

I. Critical Reading

Before beginning the investigations, read Chapter 41: Animal Nutrition in your text. You may also have to refer to other chapters in Unit Seven and Chapter 24: The Origin of Species to learn the most from this investigative case.

1. Compare the digestive systems of cattle (in this instance a cow), horses, and humans using Figure 7.2, as well as Figures 41.20 and 41.10 in your text.

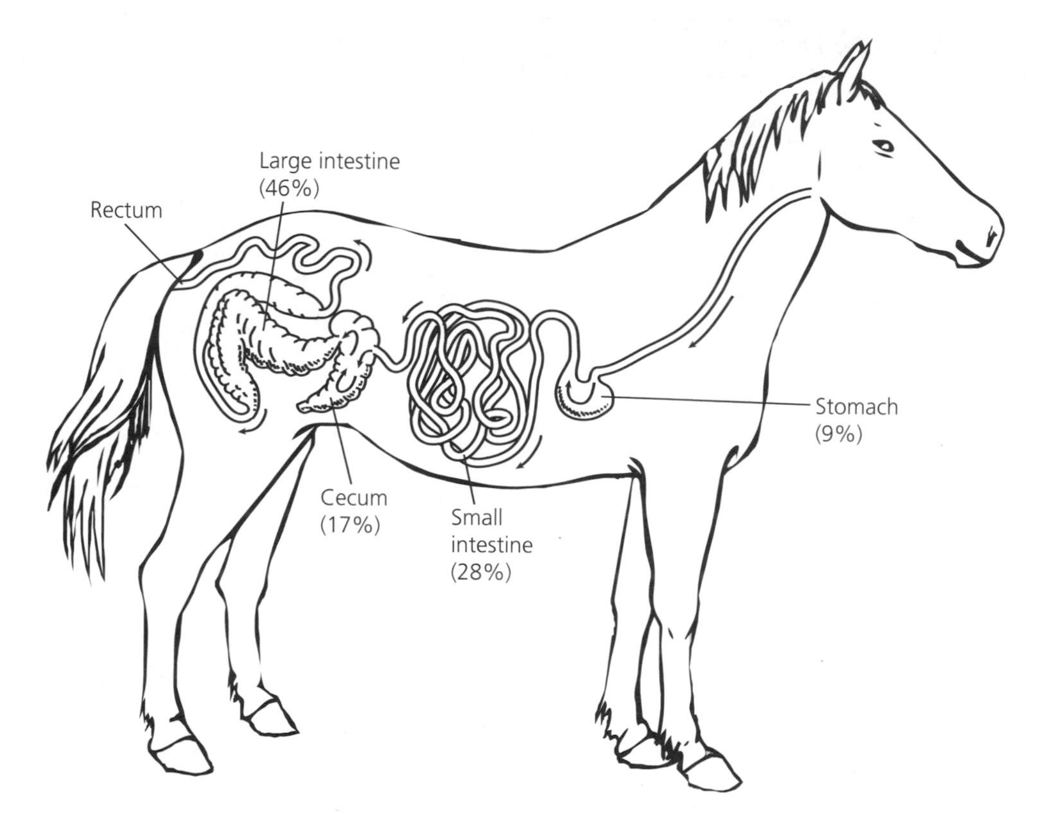

Figure 7.2 Idealized digestive system of the horse—a postgastric fermenter. The percentages refer to the relative capacities of each structure.

a. How does the structure of the stomach in these three organisms differ?

b. Examine Figure 7.3. Note the significant difference in the relative capacity of the stomach when comparing a horse with a cow. What do you think this indicates about the location of digestion in both animals?

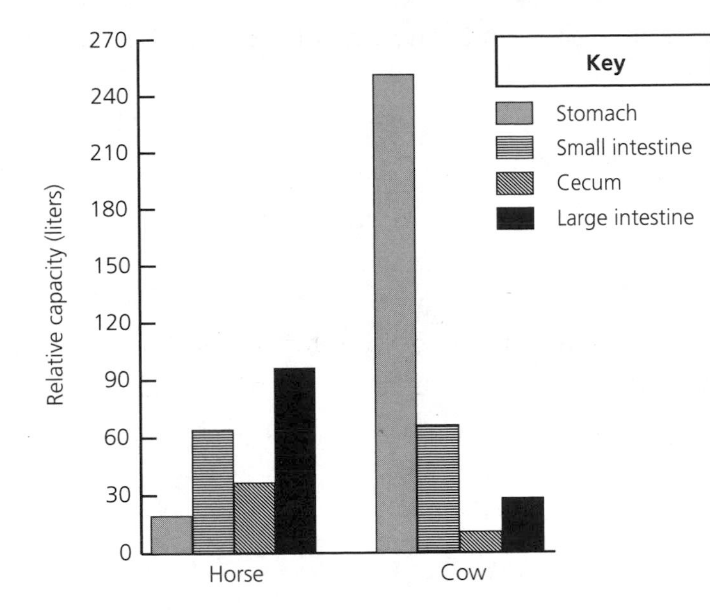

Figure 7.3 Relative capacities of the digestive tracts in horses and cows. (Adapted from Ensminger and Olentine, 1978)

c. How do you think the relative capacities of the stomach, small intestine, and large intestine would differ in humans as compared to horses and cows?

2. Where are the microbes (mostly facultatively anaerobic) involved in digestion primarily located in the cow, horse, and human?

3. Compare the function of the stomach in the cow, horse, and human. Also, comment on the ability of each organism to digest cellulose in the stomach.

4. The function and structure of the digestive tracts of the cow, horse, and human fit the diets of these animals. The grazers eat primarily leafy vegetation. Omnivores, however, consume more calories from plant storage structures (grains, tubers, and roots) than from

leaves. Compare the relative sizes and functions of the digestive systems in grazers and omnivores.

5. Once a horse swallows a bolus of food and it enters the stomach, strong muscles at the cardiac end of the stomach prevent the animal from regurgitating its food. How does this differ from the swallowing process in ruminants?

6. The digestion of what type of macromolecule begins in the organism's stomach (the abomasum in ruminants)?

7. The small intestine functions similarly in all three organisms. Describe its function.

8. The hindgut (structure of the digestive tract after the small intestine) in all three organisms contains a diverse population of fermenting microbes. These microbes release gases, as well as fatty acids and lactic acid, as waste products of fermentation. Of the three organisms, the most hindgut fermentation occurs in horses.

   If a horse feeds on too much starchy grain, a lot of undigested carbohydrate will pass from the foregut into the hindgut. The same digestion problem happens when a horse feeds on spring grasses in the pasture. Grass plants produce many carbohydrate-rich leaves during the spring, but in the summer they produce more stem than leaf. Microbes that ferment the excess starch produce an increased level of lactic acid, which lowers the pH of the hindgut. Digestion in the hindgut stops and the horse often has to be treated for impaction. Why do you suppose excess starch halts digestion?

9. The ecology of digestion relies on the presence of the right balance of microbial populations. For example, over a long period of time, antibiotic treatments can affect digestion in humans, and there are similar effects in cows and horses. Describe the effect of antibiotics and the consequences to digestion.

10. Cows, horses, and humans all consume foods containing cellulose. However, none of these organisms produces enzymes that can break down cellulose. Fermenting microbes in the stomach of cows convert cellulose to energy-rich fatty acids. Explain the fate of cellulose in horses and humans.

II. Feeding Horses

Wild horses are grazers that feed freely in grasslands. Their diet consists mainly of grasses and forbs (nongrass flowering plants). Domesticated horses are provided fewer choices. They are given access to either grass forage or hay that provides all or most of their required nutrients. To offset deficits in the quality and quantity of grass forage and hay, supplemental feed is provided. Often this feed contains additional vitamins and minerals and specialty feeds such as oats or corn.

1. Grass forage and hay consist largely of structural carbohydrates in the form of fiber, primarily cellulose. The fiber provides the raw material necessary for the growth and maintenance of microbial flora required for good horse health. We know that fiber is processed mainly in the hindgut. However, where does the processing of the fiber begin?

2. When the yearling in the case showed signs of low weight gain, he was removed from the pasture. The owners increased his total amount of feed and provided a higher proportion of supplemental feed. The supplemental feed consisted of a mixture of lysine-rich field peas and cereal grains containing methionine and cystine. Why do you think the owners decided to incorporate more amino acids in the yearling's diet?

3. Weight status and activity levels can be used to estimate the amount of feed needed by individual horses (Figure 7.4). According to this chart, should you feed a horse with a normal weight of 1,150 pounds that engages in hard activity more total feed than a horse with a normal weight of 1,150 pounds that engages in light activity?

Calculating Daily Feed for Horses				
Weight 1,150 lb				
Activity	**Status**	**Total Feed (lb)**	**Grass Forage**	**Supplemental Feed**
Light	Normal	23.00	17.25	5.75
Moderate	Normal	23.00	13.80	9.20
Hard	Normal	23.00	11.50	11.50
Light	Underweight	28.75	21.56	7.19
Moderate	Underweight	28.75	17.25	11.50
Hard	Underweight	28.75	14.38	14.38

Figure 7.4 Spreadsheet for determining daily feed mixes for normal and underweight horses based on a weight of 1,150 pounds.

4. Compare the proportions of grass forage and supplemental feed in the total feed mix for the normal-weight horse with hard activity and the normal-weight horse with light activity.

5. If you have an underweight horse that is worked hard and a horse of normal weight that is worked hard, which should receive the most total feed? Is there a difference in the ratio of grass forage to supplemental feed?

6. Some organisms that infect grass prove harmful to horses. When the mare in the case was pregnant, the owners had the pasture with tall fescue grass tested. Fescue can contain an endophytic fungus that produces ergot alkaloids that are chemically similar to neurotransmitters (Browning, 2003). These alkaloids can interfere with the physiological control of labor and delivery as well as the survival of the foal. Should a pregnant mare be removed from a pasture infected with this fungus and fed solely supplemental feed?

7. In the case, the crowding of buffalo in a neighbor's pasture is criticized. Overcrowding limits available forage and substantially increases the potential for parasite infection. Parasites can be picked up when animals eat the larvae of flatworms and roundworms living in their pasture feed. How do the larvae of these animal parasites get into the grass?

8. Easy-keeper horses are usually insulin resistant. Because their bodies do not respond to normal insulin levels, they maintain abnormally high levels of insulin compared with horses that are not easy keepers. Easy keeper horses tend to convert excess carbohydrates to fat and require less food to maintain their weight. Easy keepers easily become overweight on supplemental feed. How might this genetic trait be beneficial to horses and to their owners?

9. Owners of racehorses add oats to the supplemental feed to provide both extra protein for muscle repair and extra carbohydrates for the quick energy needed by these active horses. The excess starch in oats is processed in the large intestine. As you have learned, this may cause problems in horses. One proposed preventive treatment is to add a microbial amylase that functions at a wide range of pH to the supplemental feed. How could this help?

Additional Investigations

III. Reproduction in Mammals

A. Reproduction in *Equus*. After giving Leah the rest of the tour, Jim and Gina pointed out a pasture near their farm that contained a male zebra and a female donkey. Their healthy offspring, born earlier that year, was easily seen as it stood near the fence. Leah said, "Look at that weird little striped donkey. What is that?"

"We're guessing the donkey and zebra got together," Jim said.

"It's called a zonkey," Gina added.

"So does that mean a zebra is just a donkey with stripes?" Leah wondered aloud.

(See Figure 7.5 for images of these three animals.)

Horse, donkey, and zebra species all belong to the genus *Equus*. When two different species from *Equus* such as a horse and a donkey mate, they may produce hybrid offspring. (See the mule in Figure 24.4 in your text.)

(a) Male zebra who sired the zonkey

(b) The zonkey and its dam—a donkey

(c) The yearling zonkey

Figure 7.5 This yearling zonkey is an offspring of a male zebra and a female donkey.

1. Using the data in Table 7.1, determine the zonkey's parents. Explain.

Table 7.1 Diploid Chromosome Numbers in Various Equid Mammals

Equid Mammal	Total Chromosome Number
African zebra	44
Mountain zebra	32
Donkey	62
Zonkey	53

2. Although zebras and donkeys can produce offspring, do you think these two types of organisms should be considered to be the same species? Explain, using definitions of "species" and examples of prezygotic and postzygotic isolating mechanisms (see Chapter 24).

3. Examine the photos in Figure 7.5 and list three traits of the zonkey that are different from its dam.

B. Gestation in Mammals. Gestation is the period of time in which developing young are carried within the uterus. Table 7.2 shows the average gestation period for different placental mammals. (*Note:* Whereas gestation time usually is calculated as the time from fertilization to birth, some smaller placental mammals such as squirrels and minks have delayed implantation. In this case, gestation refers only to the total time of active development.)

1. Provide a hypothesis about the length of gestation with respect to the animals involved. For example, the larger the typical litter, the shorter the gestation period.

Table 7.2 Approximate Gestation Periods for Placental Mammals

Mammal	Approximate Length of Gestation	
Mouse	19 days	
Rabbit	31 days	
Dog	61–63 days	
Cat	63–65 days	
Pig	112 days	
Goat, pygmy	145 days	
Goat	150 days	
Deer	201 days	
Human	266 days	
Cow	284 days	
Bison	285 days	
Horse	330–340 days	
Giraffe	457 days	
Elephant, Asian	645 days	

2. What single data column would you add to Table 7.2 to help you test your hypothesis?

3. Choose six of the mammals in the table and look up the data for each. Enter your results in the blank column. (Don't forget to write in a column heading.)
4. a. To test your hypothesis, construct a line graph with the data from the table. What will be on the *x*-axis of the graph? What will be on the *y*-axis of the graph?

   b. Plot your graph in the space below.

   c. Interpret the graph. Are the results what you expected?

IV. A Closer Look at Horse Evolution

In "Critical Reading," some of the evolutionary adaptations found in the digestive systems of grazing animals were considered. In this section, further adaptations found in modern horses and in fossils of equids (horselike organisms) will be examined.

1. To gain an overview of over 50 million years of horse evolution, examine Figure 25.25, the branched evolution of horses (cladogenesis). At one time, horse evolution was often depicted as a straight line—implying that an accumulation of changes gradually transformed one species into a species with different characteristics (anagenesis). (See Figure 7.6.) How does the shape of the more modern phylogenetic tree contradict the impression that one equid taxon died out as the next taxon emerged?

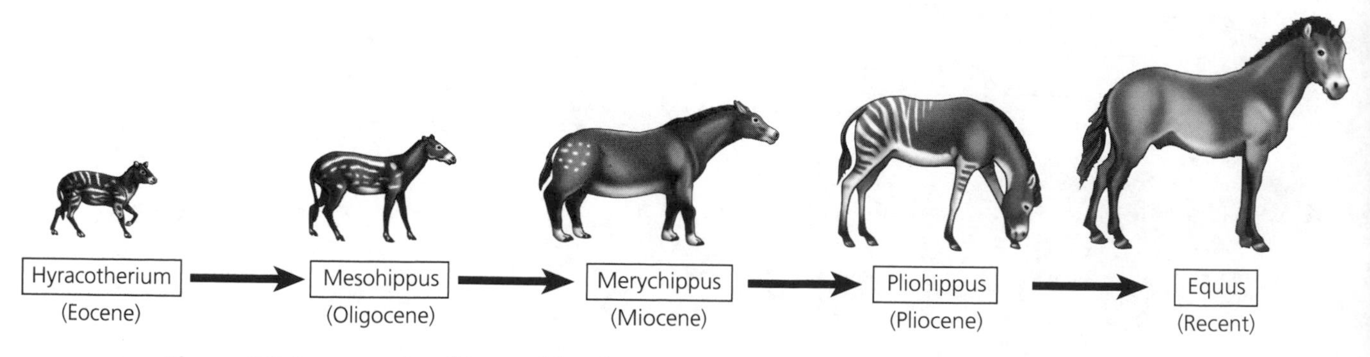

Figure 7.6 An example of how older diagrams depicted horse evolution as a straight line.

2. Go to "Wild Horses—An American Romance" at http://net.unl.edu/artsFeat/wildhorses/wh_origin/wh_origin2.html and explore the interactive phylogenetic tree of horses. Compare the tree to the time line at http://net.unl.edu/artsFeat/wildhorses/wh_origin/wh_origin.html. What information does this interactive tree and time line provide as compared to Figure 25.25?

3. Summarize the changes in feeding habits, foot structure, and overall size of the animals depicted in Figure 25.25 on the one branch leading from *Hyracotherium* to *Equus*. (Note that the trends you see on this branch did not occur together or at a steady rate. They may have occurred in several branches of this evolutionary tree at different times. Evolution is not goal oriented.)

4. Examine Figure 7.7. Compare the teeth of the extinct and extant equid jaws.

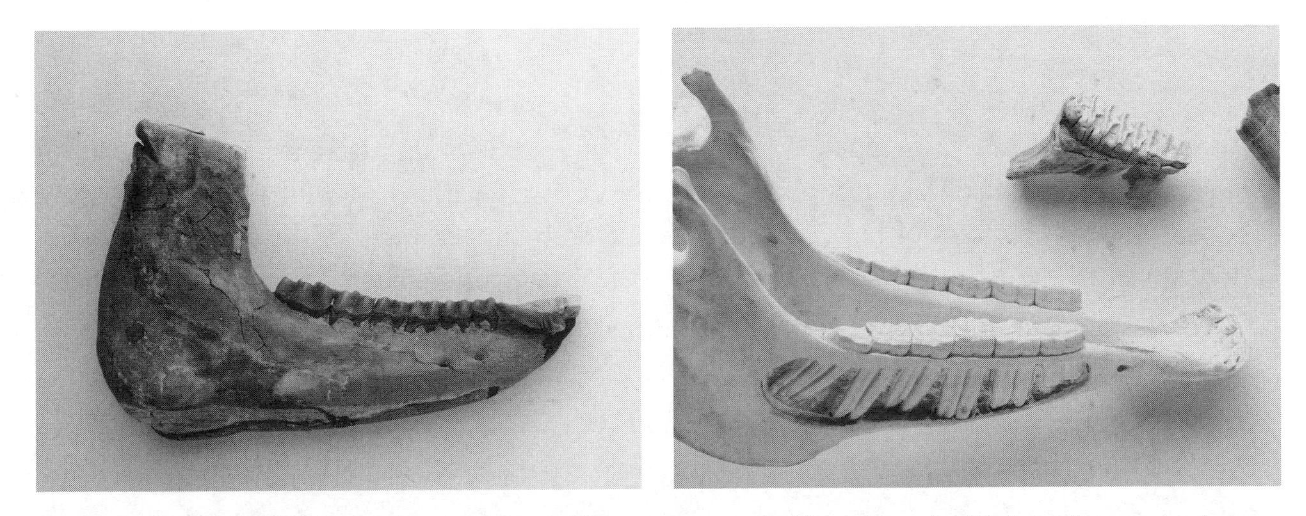

Figure 7.7 Jaws with teeth from equid of 18 million years ago (left), compared with jaws from a modern horse (right). Note that the modern jaw is much larger than the fossil jaw.
Source: Philip Dowell/Dorling Kindersley (left); Colin Keates/Dorling Kindersley, Courtesy of the Natural History Museum, London (right).

5. What kind of food did the extinct equid likely eat?

6. Although well adapted for running, modern horses are susceptible to a foot condition called laminitis, a deterioration of connective tissues within the hoof (Figure 7.8). Strong fibrous tissues called laminae occupy the space between the bone and hoof. The laminae support the terminal bone (3rd phalanx) and connect it to the hoof wall. In horses suffering from laminitis, these connective tissues become swollen and then detached, allowing the bones to twist, penetrate the hoof, or sink within the hoof. If not properly treated, the animal will become lame. In some cases the lameness can be remedied, but sometimes the lameness is so debilitating that the animal needs to be euthanized.

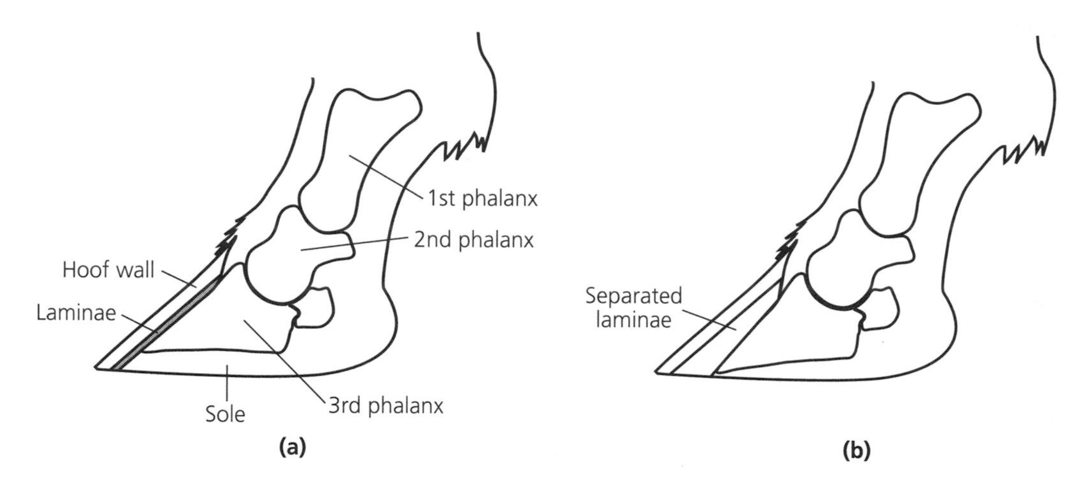

Figure 7.8 Laminitis. (a) Normal foot anatomy. (b) Foot of horse with laminitis. Note the shifting bone structure.

Although the cause of laminitis is not known, many predisposing factors have been correlated with the onset of laminitis. One of these factors is overfeeding with lush spring grasses or other sources of high carbohydrates. One hypothesis is that the excess carbohydrates lead to toxins in the blood that irritate the sensitive laminae.

What are some differences between the normal foot and the foot with laminitis? Consider where the weight of the horse is placed.

7. Raising horses is at best an artificial system in which the horse's normal movement, choice of feed, population size, and breeding are restricted. What evolutionary implications might this have for future domesticated horse populations?

V. Open-Ended Investigations

Go to the Case Book website at http://bioquest.org/icbl/casebook to access a live copy of the Excel spreadsheet on feeding horses. Investigate the scenarios provided there or make up your own.

References

Browning, R., Jr. "Tall Fescue Endophyte Toxicosis in Beef Cattle: Clinical Mode of Action and Potential Mitigation Through Cattle Genetics." 2003. http://www.bifconference.com/bif2003/BIFsymposium_pdfs/Browning.pdf

Ensminger, M. E., and C. G. Olentine. *Feeds and Nutrition*. Clovis, CA: The Ensminger Publishing Company, 1978.

Fossil Horse Cybermuseum. http://www.flmnh.ufl.edu/natsci/vertpaleo/fhc/firstCM.htm (accessed October 11, 2007).

Hunt, K. "Horse Evolution." 1995. http://www.talkorigins.org/faqs/horses/horse_evol.html

CHAPTER 8

Back to the Bay

Back to the Bay

Descended from oystermen and crabbers, Liam and Solana looked forward to their family reunion at Maryland's Chesapeake Bay. Although this section of the bay no longer supported commercial shell fishing, there was plenty to see.

While waiting in line to purchase tickets for a boat tour of the bay, Liam and Solana heard a series of screeches. They scanned the docks for the source of the mayhem. A large gull that had settled on a covered boat hurriedly flew away midway through the screeches. After several more seconds, the noise stopped.

"What *was* that?" Solana asked. "It sounds like birds are being attacked, but I don't see anything." Liam just shrugged and pointed out a mallard still floating a few feet from the dock and a tern still sitting on the nearby pilings despite the noise. "Can't they hear? They'd have to be deaf to ignore that," Solana said.

The woman working at the nearby snack stand looked up and offered an explanation. "Well, it's a tape the marina owner is using to scare the gulls away. Seems like there's more every year. Gull droppings cause holes in the boat covers if they aren't cleaned quickly. We've tried everything to keep the gulls away.

The marina owner found an ad for this tape on the Internet. He says there are six different gull distress calls, which are all supposed to sound just like the real thing. The tape plays every couple of hours during the day. It worked great last summer, but lately some of the gulls seem to ignore it.

"The owner of the next marina over even keeps a few dead gulls out on the dock. He claims it keeps the tapes working, but I . . ."

"Look at that," Liam interrupted. "Someone used duct tape to block the speaker at this end of the dock."

Squinting her eyes to get a better look at the speaker, the woman nodded. "I bet the guy in the 22-foot Sea Ray did it," she offered. "He claims that it's the new landfill causing the gull problem in the first place, and we should get rid of that rather than disturb his peace."

"Well, at least the tape still seems to be working on him," observed Liam, laughing.

Figure 8.1 Gull populations often become a nuisance to people in areas such as marinas.

CASE ANALYSIS

1. **Recognize potential issues and major topics in the case.** What is this case about? Underline and list terms or phrases that seem to be important to understanding this case. Then list **3–4** biology-related topics or issues in the case.

2. **What specific questions do you have about these topics?** By yourself, or better yet, in a group, list the things you already know about this case in the "What Do I Know?" column. List questions you would like to learn more about in the "What Do I Need to Know?" column.

What Do I Know?	What Do I Need to Know?

3. Put a check mark by **1–3** questions or issues from the "What Do I Need to Know?" list that you think are most important to explore.

4. **What kinds of references or resources would help you answer or explore these questions?** Identify two different resources and explain what information each resource is likely to give that will help you answer the question(s). Choose specific resources.

Core Investigations

I. Critical Reading

Read Chapter 51: Animal Behavior, which discusses several kinds of animal behavior. Listed below are descriptions of various responses displayed by birds in the case. Answer the questions that follow each response. Provide examples from Chapter 51 to support your answers. As you complete this exercise, note that, behaviorally speaking, distress calls are similar to the alarm calls you read about in the text.

1. Only one species of gull leaves the area when the tape plays.
   a. Explain the gulls' response.

   b. Do you think this response is primarily learned or primarily genetic? Explain.

   c. Could the call be categorized as a signal? Explain.

2. A year later, the same species of gull remains when the tape plays.
   a. What might account for the change in the gulls' response?

   b. Do you think this response is primarily learned or primarily genetic? Explain.

   c. Speculate on what kinds of behavior this response might be.

   d. How does this behavior exemplify the "cry wolf" effect? Explain this in terms of a cost-benefit analysis.

3. The same species of gull resumes its response to the tape when dead gulls are displayed in the immediate area of the sound.

a. What might account for the change in the gulls' response?

b. Do you think this response is primarily learned or primarily genetic? Explain.

c. How does this behavior reflect associative learning? Is it negative or positive reinforcement?

d. Can this behavior be interpreted as altruistic? Explain.

4. Birds other than this gull species ignore the taped distress calls.

a. What might account for the observed behavior in the other birds?

b. Researchers switched young of one species of bird with those of another. The young birds responded to the signals of the new species. Do you think this response is primarily learned or primarily genetic? Explain.

c. What evidence do you have from the case to support the idea that calls are species specific?

5. You observe that some gulls continue to respond to taped distress calls when *no* danger is present. Do you think that these gulls are more or less fit than those that stop responding to such calls? Explain.

II. Design an Experiment

Design an experiment to examine the following question. Is the response to distress calls (moving away from the area) learned in herring gulls?

Materials: You will have access to newly hatched, juvenile, and adult herring gulls. You also will have a tape of herring gull distress calls.

The following additional resources may help you with this activity:

Web/CD Chapter 51 Investigation, How Can Pillbug Responses to the Environment Be Tested? *and Lab Topic 1 of* Investigating Biology *by Morgan and Carter.*

1. Restate the question being studied as a testable hypothesis.

2. Describe the experiment.
   a. What will the treatment(s) be? Which animals will receive each treatment?

   b. What will you measure as the response to the treatment?

3. List three variables that you will control.

4. Describe the experimental results that would support your hypothesis.

III. Biology in Advertising

Examine the advertisement for a gull distress call recording in Figure 8.2 and answer the following questions.

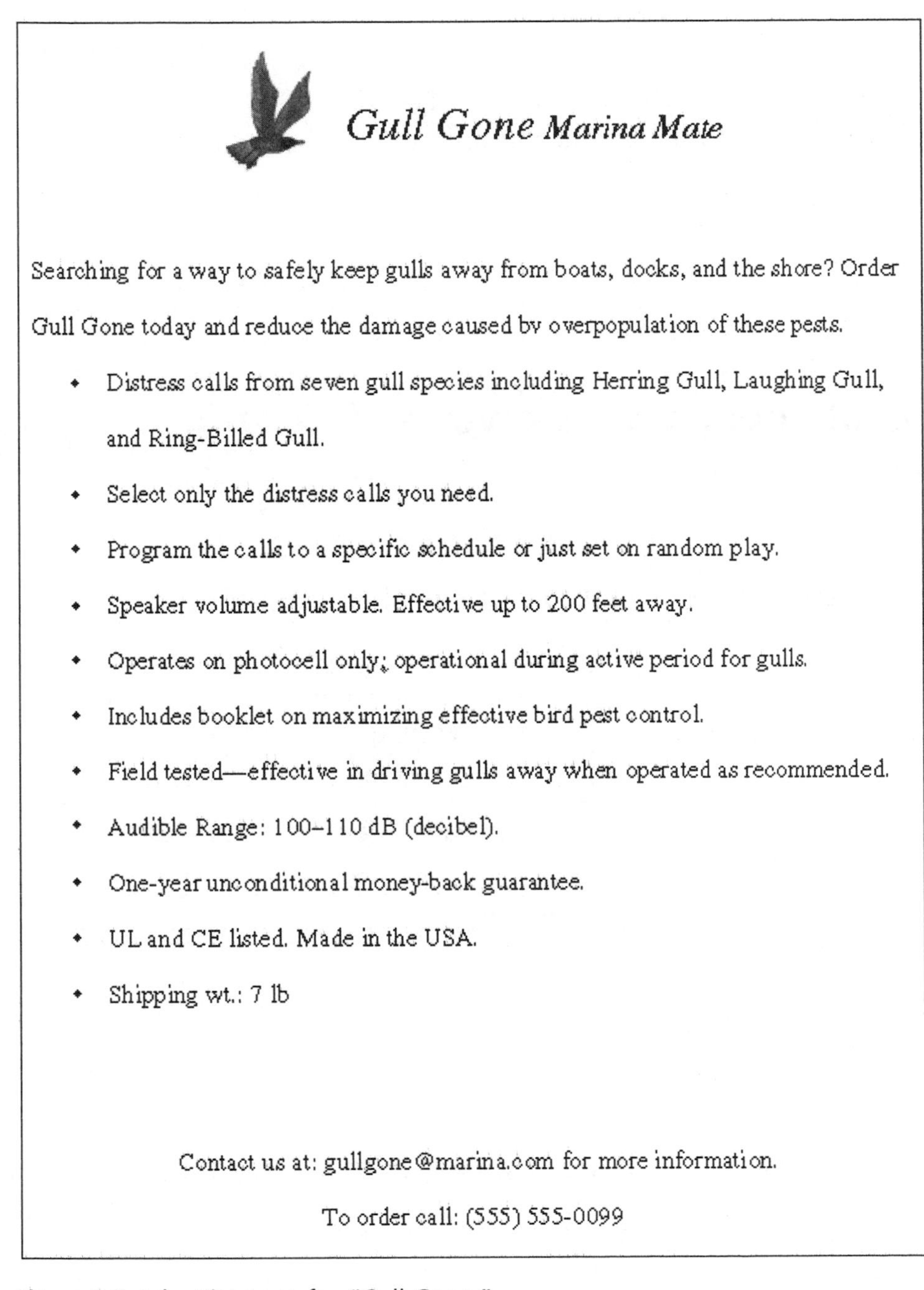

Figure 8.2 Advertisement for "Gull Gone."

1. What are three biological claims in this advertisement?

2. Choose one of the claims and briefly describe an experiment that would test its validity.

3. Is there any behavioral significance to the product's feature that allows for playing distress calls in random sequence?

IV. Investigations of Population Growth and Control

A. Gull Population Growth *(Refer to Chapter 53 in your text for help with this exercise.)*

1. Review the data in Table 8.1.

Table 8.1 Increase in Number of Gull Mating Pairs in Selected Locations

Locations	Initial Observation Mating Pairs/Year	Final Observation Mating Pairs/Year	Gull Species
Kennedy Airport, New York, United States	15 pairs/1970	7,600 pairs/1990	Laughing Gull
Leslie Spit, Toronto, Canada	20 pairs/1973	80,000 pairs/1982	Ring-Billed Gull
Five Islands, Wollongong, New South Wales, Australia	3 pairs (est.)/1949	51,500 pairs/1978	Silver Gull
Mud Islands, Port Phillips Bay, Victoria, Australia	5 pairs/1959	70,000 pairs/1988	Silver Gull

a. Are the four gull populations all increasing in size at the same rate? Explain.

b. Which rates are the most similar?

c. While doing further research on gull populations, you discover that in 1985 there were 15,000 pairs of laughing gulls living in the vicinity of John F. Kennedy International Airport in New York City. How does this knowledge change your answers to 1.a. and 1.b.?

2. Consider reasons why differences in these gull populations occur. List three ways that the environment in which the birds live could affect their rates of reproduction.

3. Population growth is greatly influenced by environmental factors. The activities of a human population impact nearby gull populations. For example, the potential for collisions between aircraft and gulls ensures that officials at John F. Kennedy International Airport implement gull population control measures. List at least three other examples of human–gull interactions. For each example, explain whether the gull population benefits.

B. Control of Bird Populations: The Chicken or the Egg? If the ultimate goal at an airport location is to reduce gull population size to ensure human safety, should gull nests and eggs or adult gulls be removed? In this exercise, you will use a model to compare the effectiveness of these two methods.

1. Fill in the worksheet in Figure 8.3 to estimate the relative effectiveness of these two different population control methods.

The Chicken or the Egg Worksheet				
Find the differences in population growth resulting from two gull control methods.				
Assumptions				
Basic Population	**Enter**	**Method: No Control**	**Remove Adult Birds***	**Remove Nests and Eggs**
Adults = (2) (#nesting pairs) =	20	Nesting Pairs = 10	Nesting Pairs = 10	Nesting Pairs = 10
Total # eggs = (#eggs per nest) (#nesting pairs) =	30	Eggs per Nest = 3	Eggs per Nest = 3	Eggs per Nest = 3
Potential Population =	50	Survival of Adults = 90%	Survival of Adults =50%	Survival of Adults =90%
		Survival of Young Birds = 50%	Survival of Young Birds = 50%	Survival of Young Birds = 10%
Enter the values from the information above and perform the calculations.		**No Control**	**Remove Adult Birds**	**Remove Nests and Eggs**
Surviving Adults = (% survival) (#nesting pairs)(2)		(.90)(10)(2) = 18		
Surviving Chicks =(% survival)(#eggs per nest)(#nesting pairs)		(.50)(3)(10) = 15		
Surviving Population = Surviving Adults + Surviving Chicks		18 + 15 = 33		

*Note that the percentage of surviving young birds does not change when adult birds are removed because the adults are removed before they reproduce.

Figure 8.3 The Chicken or the Egg Worksheet. Use the provided equations to calculate surviving gull populations after implementing two types of gull control methods. This spreadsheet also is provided on the Case Book website as a functioning model. (Weisstein, 2004a)

2. Which of these gull control methods is more effective?

3. Which of these two control methods would you advocate? Explain your choice.

V. Explore the Environmental Conditions of the Chesapeake Bay

Liam and Solana spent 4 hours on their boat tour of the Chesapeake Bay, learning about its history, ecology, and geology. In this investigation, you will take a brief "armchair tour" of the Chesapeake Bay by exploring some of the environmental factors that affect its biological diversity. Go to the Case Book website for resources on Chesapeake Bay and further directions for the following activities.

A. Stressors on the Bay. Examine the animation. List three important stressors on the Chesapeake (see the Stressors on the Bay link on the Case Book website).

B. Cutter Marina. Examine the information about Cutter Marina, including the data charts and interpretation of each of the water quality variables measured. Describe and provide the values for three variables that indicate a healthy bay.

C. Eyes on the Bay. The water quality of the Chesapeake is sampled daily at more than 100 sampling stations. These data are reported and compiled online at "Eyes on the Bay" (Figure 8.4) whose link is listed on the Case Book website.

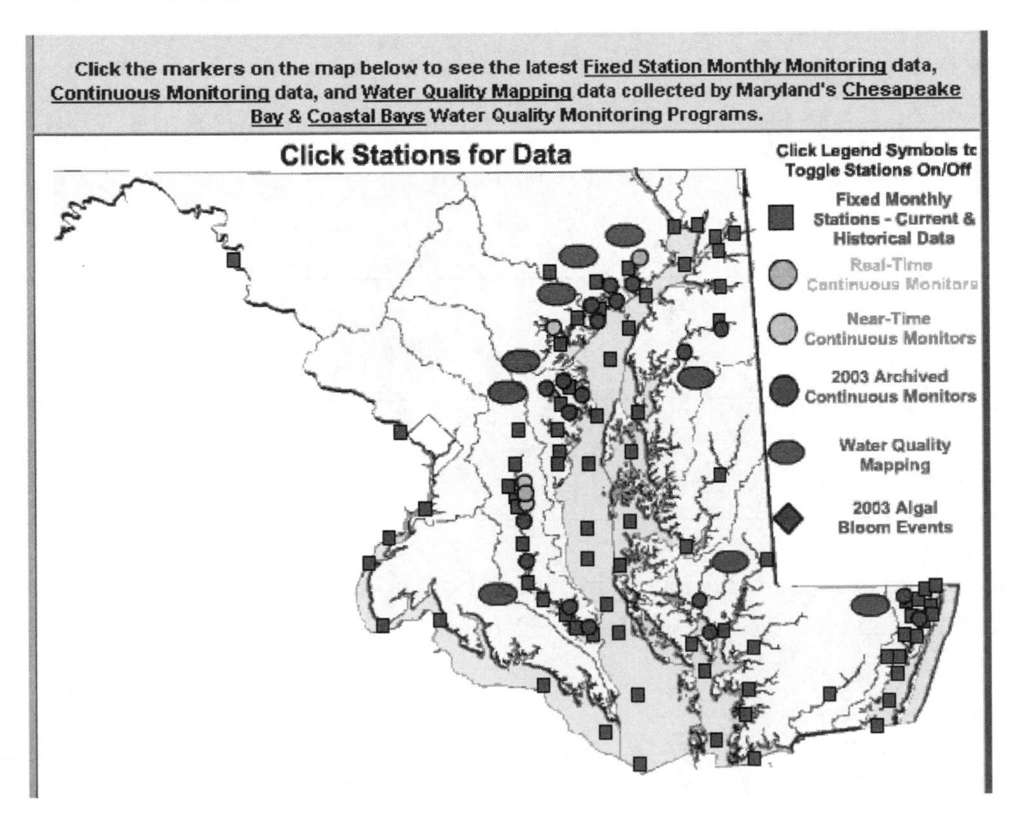

Figure 8.4 Eyes on the Bay.

1. At the Eyes on the Bay website, examine differences in salinity across the Chesapeake Bay by first switching to "full-screen map" and then running your cursor over the red square icons and reading the salinity data that appear on the left side of the screen.

a. Describe what happens to salinity as you move from open ocean (the right and lower edges of the map) to the top of the Chesapeake Bay.

b. Describe what happens to salinity as you move from the center of the Chesapeake Bay and up the Potomac River, which starts near the bottom center of the map.

2. Choose data from two stations, one from the upper Chesapeake Bay and one from the lower Chesapeake Bay near the ocean. Click on the icon to open the station's site. Look at the following variables during August: dissolved oxygen, water temperature, pH, and water clarity. Enter the data for each station in Table 8.2.

Table 8.2 Comparison of Data from Two Stations

Year ________

Month: August

Lower Bay station name:	Upper Bay station name:
Dissolved oxygen	Dissolved oxygen
Water temperature	Water temperature
pH	pH
Water clarity	Water clarity

3. Studies have shown there are major differences in types of organisms inhabiting different areas of the bay. How do the data you recorded in Table 8.2 help to explain this finding?

D. Chesapeake Bay Remote Sensing Program. Go to the Case Book website to access the link to the "Chesapeake Bay Remote Sensing Program" website (Figure 8.5). Use this website to examine productivity in the Chesapeake as measured by chlorophyll production.

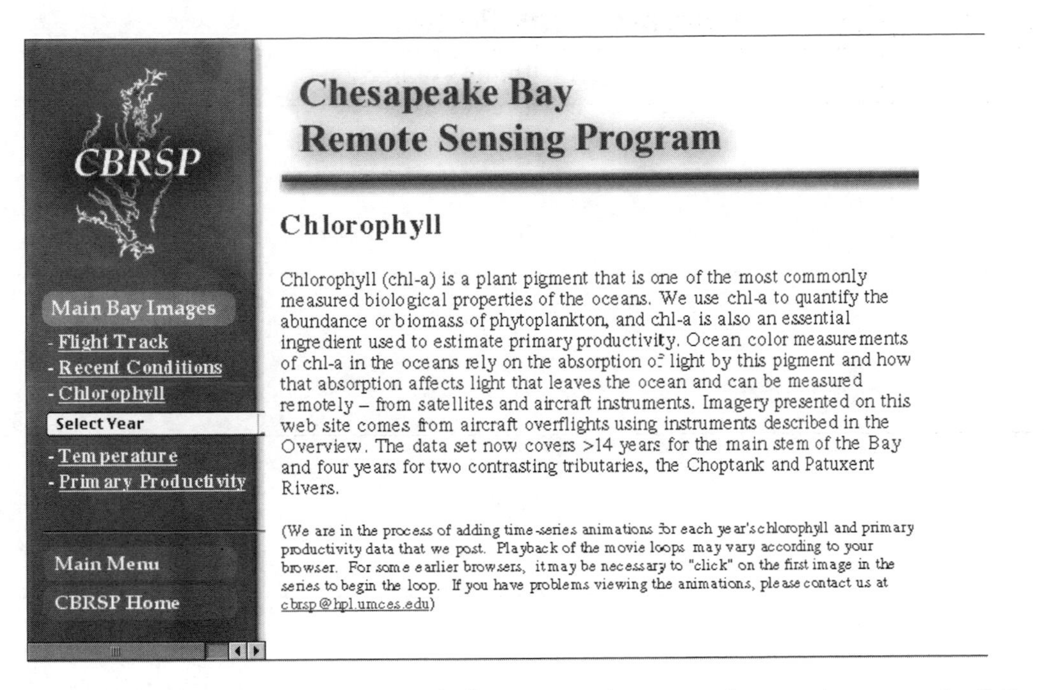

Figure 8.5 CBRSP ascertains chlorophyll concentrations to estimate primary productivity and to gauge the overall productivity of the Chesapeake Bay ecosystem.

1. Choose a year from 1998 to 2002 and run the animation of chlorophyll for the whole year. (*Note:* If you click on the animation while it is running, it will pause on the map shown. Double-clicking will resume the animation.) Briefly describe the changes in chlorophyll content in April, July, and October.

2. For the years 1998–2002, track chlorophyll data for the last reading in July. Focus on the grid square bounded by 37–37.5 (representing latitude) on the left and by 76.5–76 (representing longitude) on the bottom. Give a range and an average for each date in Table 8.3.

Table 8.3 Chlorophyll Data over a 5-Year Period

Date	Range/Average
July 1998	
July 1999	
July 2000	
July 2001	
July 2002	

3. Compare the data you tracked for 1998 and 1999. What might account for the differences you see in the chlorophyll increase for these 2 years?

4. The area of the Chesapeake Bay where Liam and Solana's family has lived is near the Lower Chester River, south of Eastern Neck Island (38.99 latitude, 76.22 longitude). Describe one variable that you think may be contributing to the failure of this part of the bay to support commercial shell fishing.

5. Summarize what you have learned about the environment of the Chesapeake Bay, comparing and contrasting the upper and lower regions of the bay and the rivers that empty into the bay.

Additional Investigations

VI. Ethics Paper on Gull and Human Interactions (Choose A or B.)

Go to the Case Book website for links to information about many of these topics.

A. Choose one of the following situations in which humans and gulls interact.

- Landfills
- Airports
- Offshore oil rigs
- Roof-roosting gulls in Melbourne, Australia
- Municipal reservoirs
- Maintaining biodiversity in wildlife refuges
- Maintaining biodiversity and gull effects
- Protecting birds from oil spills using sound

Research the situation examining control methods used and outcomes. Consider the impact of the large gull populations on the environment, other species, and humans. Then consider the impact of the control measures. Discuss the ethical issues involved in the decision to control a species and in the various control measures that were used.

B. Choose three of the following methods of gull control. Evaluate each for its effectiveness in controlling gulls, its expense, and the ethics of both using and selling these methods. Describe your findings in a 1- to 2-page paper.

- Model aircraft (Melbourne area)
- Owl effigies (Louisiana Land and Exploration)
- Rubber snakes (Tides Stadium, 1997)
- Avipoint installations to prevent roosting
- Raptors as gull predators
- Deterrent spray
- High-pressure water blaster (Melbourne area zoo)
- Monofilament lines (Melbourne area hotel)

VII. Open-Ended Investigations

You may explore gull population and control using either the "logistic growth" or "control and cost" Excel models (Weisstein 2004b and 2004c) at http://bioquest.org/icbl/casebook/gullcontrol.

References

Chlorophyll animation. http://www.cbrsp.org/cbrsp_toc_mb_chl_page.htm. Accessed October 11, 2007.

Eyes on the Bay. http://mddnr.chesapeakebay.net/eyesonthebay/index.cfm. Accessed October 11, 2007.

Weisstein, A. 2004a. Chicken or the Egg: Excel worksheet. All Weisstein models are at http://bioquest.org/icbl/casebook/gullcontrol

Weisstein, A. 2004b. Controlling Cost: Excel worksheet.

Weisstein, A. 2004c. Logistic Growth: Excel worksheet.

CHAPTER 9

Pandemic Flu (Past and Possible)

Pandemic Flu (Past): An Oral History by Teamus Bartley, Recorded by Nyoka Hawkins, 1987

T. BARTLEY: . . . [Y]ou ain't old enough to remember the year the flu struck the people so bad in this . . . in this country, do you?

HAWKINS: No.

T. BARTLEY: That was in 1918.

HAWKINS: Yeah, I think that both my . . . both my great-grandparents died in that.

T. BARTLEY: . . . [Y]eah, they did. Yeah. It was the saddest looking time then that ever you saw in your life. My brother lived over here in the camps then and I was working over there and . . . I was dropping cars under the tipple. And the fl- . . . that . . . that . . . epidemic broke out and people went to dying and it was just four and five every night dying right there in the camps. Every night. And I began going over there. My brother and hi- . . . all his family took down with it. It . . . what did they call it? The flu? Yeah. Eight- . . . 1918 flu. And when I'd get over there, I'd ride my horse and go over there of a evening, I'd stay with my brother about three hours and do what I could to help them and every one of them was in the bed and sometimes Dr. Preston would come while I was there. He was the doctor. And he said, "I'm trying to save their lives, but I'm afraid I ain't going to." They was so bad off.

HAWKINS: __________

T. BARTLEY: And every . . . nearly every porch . . . every porch that I'd look at had . . . would have a casket box setting on it. And men digging graves just as hard as they could and the mines had to shut down. There wadn't nary a man . . . there wadn't a . . . there wadn't a mine running or a lump of coal running nowhere. . . .

T. BARTLEY: Stayed that a way for about six weeks.

Figure 9.1 Mining family in Pennsylvania near the time of the 1918 epidemic. From the Stanley family records, circa 1919, privately held.

CASE ANALYSIS

1. **Recognize potential issues and major topics in the case.** What is this case about? Underline terms or phrases that seem to be important to understanding this case. Then list 3–4 biology-related topics or issues in the case.

2. **What specific questions do you have about these topics?** By yourself, or better yet, in a group, list what you already know that is related to the case in the "What Do I Know?" column. List questions you would like to learn more about in the "What Do I Need to Know?" column.

What Do I Know?	What Do I Need to Know?

3. Put a check mark by **1–3** questions or issues in the "What Do I Need to Know?" list that you think are most important to explore.
4. **What kinds of references or resources would help you answer or explore these questions?** Identify two different resources and explain what information each resource is likely to give that will help you answer the question(s). Choose specific resources.

Core Investigations

I. Exploring Flu Antigens, Genetics, and Replication

The transcript in the case is an oral history told by a survivor of the deadly flu pandemic (global epidemic) of 1918. It is estimated that this flu was responsible for at least 40 million deaths worldwide. Questions about the 1918 flu and why it was so deadly were unanswered for years. In 1995, scientists successfully sequenced the genome of the virus using archival autopsy tissues of known victims from 1918 (Taubenberger and Morens, 2006). The virulent human-to-human transmissible influenza of 1918 is an avian flu virus that scientists named H1N1. Less lethal descendants of H1N1 are among those causing the seasonal flu for which we get immunizations (vaccinations) each year.

This investigation emphasizes Concepts 19.1, 19.2 (read Reproductive Cycles of Animal Viruses), 19.3 (read Viral Diseases in Animals and Emerging Viruses), and Table 19.1. Be sure to examine the diagram of the flu virus.

Both H1N1 from 1918 and today's closely monitored avian flu virus, H5N1, are classified as influenza A viruses.

1. What molecule does the H stand for in the name of both viruses; where is the H molecule located on the virus; and what role does the H molecule play in the life cycle of the virus?

2. What molecule does the N stand for, where is it located; and what is its function in the viral life cycle?

The internal structure of influenza A flu viruses includes a matrix surrounding eight individual negative-stranded RNA molecules, each within its own capsid, that make up its viral genome. Each of the RNA strands codes for a specific protein.

3. The virus uses the host cell to produce more copies of itself. As the viral RNA replicates in the host, mutations arise much more frequently in the viral genome than in the host genome (which replicates DNA). Why do more mutations occur in the virus than in the host?

In addition to mutation (antigenic drift), influenza A viruses can also exchange RNA with other strains of influenza A that may have simultaneously infected the same host cell. The RNA molecules can undergo genetic recombination, producing new strains with unique viral genomes in a process called antigenic shift. New strains of influenza A are produced continuously through both mutation and recombination, necessitating annual seasonal flu shots to prevent infection.

4. Depending on the strain, influenza A is found in a variety of animal hosts including humans, pigs, birds, cats, dogs, and even whales. Many influenza A viruses are not specific for a single kind of animal host. In the later discussion, "Pandemic Flu (Possible)," maps showing the current spread of H5N1 include the density of chicken farms and pig farms as well as human populations. Why do you think these three populations are being tracked by epidemiologists looking for new strains of avian flu?

5. Suppose a human host suffering from seasonal influenza A (readily transmitted from human to human) comes down with avian flu at the same time. What serious consequences might result?

6. The immune system responds specifically to the exact combination of the H and N surface proteins found in a flu virus particle. Currently, 16 antigenically distinct groups of H molecules and 9 distinct groups of N molecules have been identified. How many possible antigenically distinct combinations of H and N could occur?

Fortunately, not every combination of H and N proteins causes the flu in human hosts. In addition, not all influenza is human-to-human transmissible like the deadly H1N1 flu of 1918. In fact, as of April 2007, the only known human-to-human transmissible influenza A strains include the H1, H2, or H3 proteins combined with the N1 or N2 proteins. These strains are transmitted via virus-laden droplets that are coughed or sneezed into the air or onto surfaces and then contact susceptible tissues.

Other combinations of the H and N proteins cause flu in other species, but they cannot easily initiate disease in humans. The H5N1 "bird" flu is deadly, but it was not known to be human-to-human transmissible at the time this book was published.

When a viral hemagglutinin attaches to a specific receptor on a host cell, it initiates the process by which the virus gains entry into the host cell. Without these receptors, infection cannot occur.

See Table 9.1 for differences in the receptors for H1, H2, and H3 versus H5.

Table 9.1 Comparing Hemagglutinin Receptors in Humans and Birds

H1, H2, and H3: Human Flu	H5: Avian Flu
Receptors are host cell–surface glycoproteins that terminate in sialic acid linked to the penultimate (next to last) galactose by an alpha 2,6 linkage.	Receptors are host cell–surface glycoproteins whose terminal sialic acid is linked to the penultimate galactose by an alpha 2,3 linkage.
Cells with these glycoproteins are common in the human upper respiratory tract (nose, upper throat).	Cells with these glycoproteins are common in bird upper respiratory tracts.
	Human cells having glycoproteins with the alpha 2,3 linkages are found deeper in the human respiratory tract, throat, and deeper in the lungs.

Table based on information in Kuiken et al., 2006.

7. **a.** Explain how the location of human glycoprotein receptors for H1, H2, and H3 influenza A antigens facilitates transmission of these strains among humans.

   b. Explain how the location of human glycoprotein receptors for H5 influenza A antigens has contributed to the failure—so far—of this virus to be readily transmitted among humans.

8. The World Health Organization Global Influenza Programme is developing "pre-pandemic" candidate vaccines based on H5N1 viruses from infected humans. These pre-pandemic vaccines are needed for preparedness planning and are used in studies that inform dosage, cross-reactivity, and cross-protection. They are also available for governments to request for pilot vaccine production (World Health Organization, 2007).

   Do you think that these pre-pandemic vaccines will be effective against an H5N1 influenza that becomes human-to-human transmissible? Why or why not?

9. A new strain of H5N1 avian flu could emerge that would result in human-to-human transmission.

   a. Why is vaccine development not the highest priority right now?

   b. Once an epidemic has started, vaccine development will be a high priority. Epidemics often occur in "waves" of illness, with different segments of the population becoming ill each time. Three waves of disease, each lasting weeks, are not rare in a flu epidemic. Why does creating a vaccine in preparation for a second wave of an epidemic make the most sense?

10. Review the replication cycle for flu viruses in Chapter 19. Pharmaceutical companies wish to produce antiviral drug therapies that can interrupt the viral replication cycle while causing minimal adverse effects on the host cells. Several classes of drugs are known to interrupt influenza A replication.

Oseltamivir (Tamiflu) and zanamivir (Relenza) are drugs that inhibit the action of the N protein. Oseltamivir has been used successfully in some cases but only during the first 36–48 hours postexposure. What part of viral replication is affected by these two drugs? Why is early treatment needed?

11. Resistance of avian influenza A to oseltamivir has already been observed. For example, a resistant H5N1 influenza A mutant has been isolated from a human. There was a substitution of a single amino acid at position 274, in which a tyrosine replaces histidine (deJong et al., 2005). Explain how resistance to a drug increases the fitness of the virus.

II. Pandemic Flu (Possible): The Spread of H5N1 Avian Influenza

Although the 1918 "Spanish flu" pandemic is long over, its lessons have been critical to understanding subsequent flu pandemics (in 1957 and 1968) as well as for preparing for future pandemics. The current H5N1 avian influenza is a different type of flu virus than those that have infected humans in the past. At the time of this publication, H5N1 has been only avian-to-human transmissible, but it is frequently deadly in the people who contract it from birds. In this investigation, you will be making observations from the maps that follow (Figure 9.2), detecting patterns, and drawing inferences about the occurrence and spread of H5N1 avian influenza. Use the maps to learn more about the H5N1 avian influenza A.

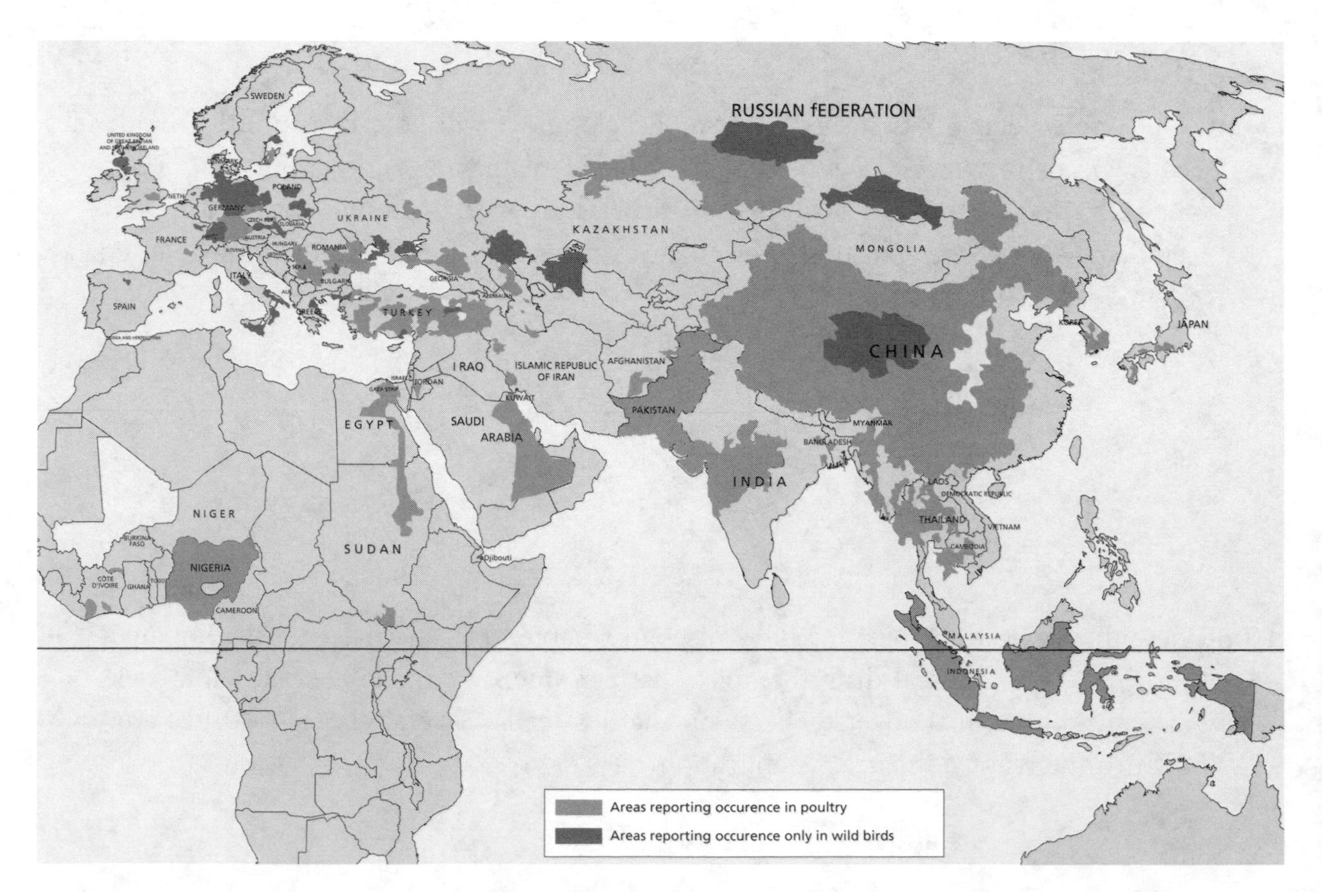

Figure 9.2a Confirmed occurrence of H5N1 avian influenza in poultry and wild birds, March 2003–2007. The horizontal black line is the approximate location of the equator. (*Source:* Adapted from World Health Organization. © WHO 2007. All rights reserved.)

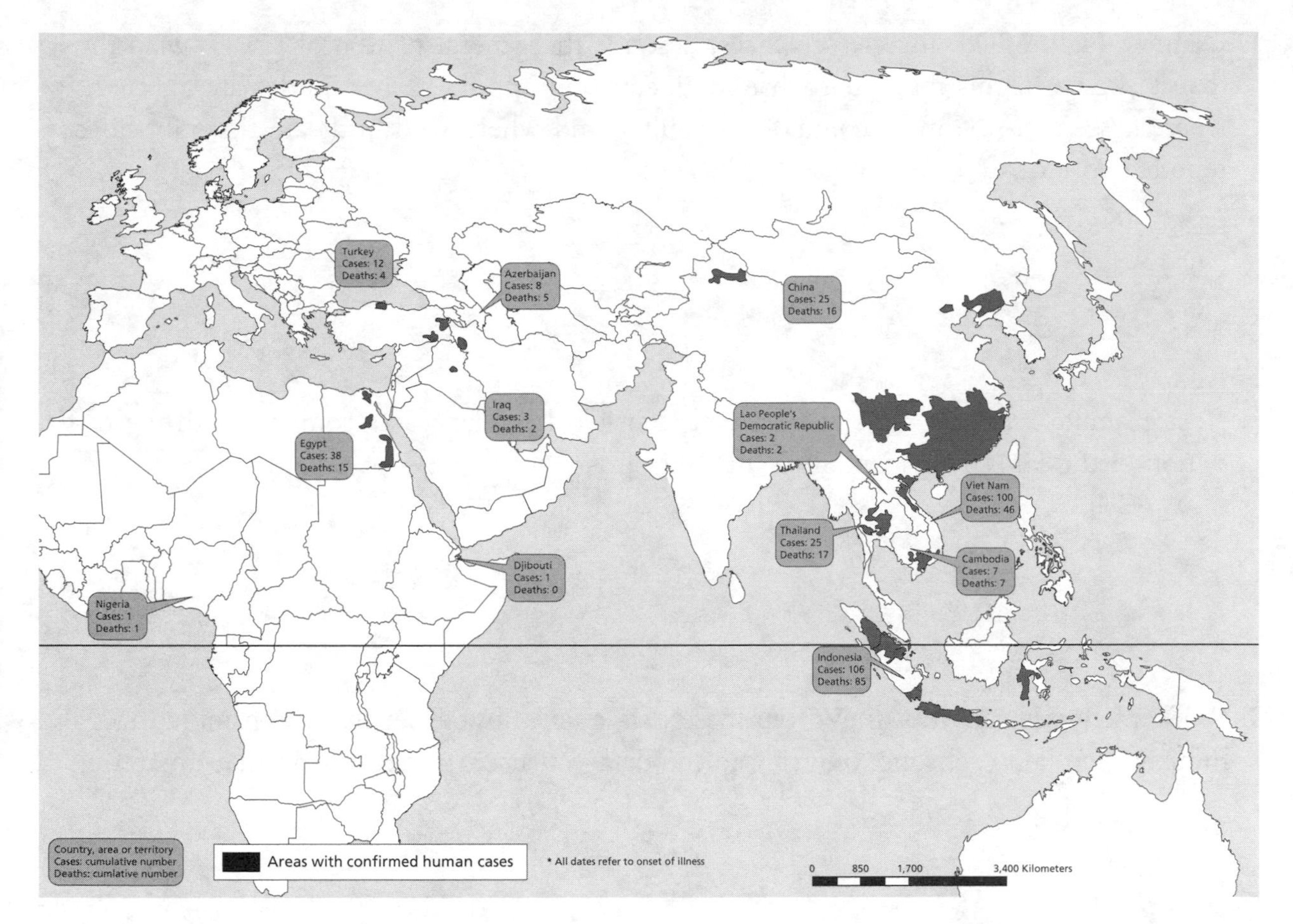

Figure 9.2b Areas with confirmed human cases of avian influenza, 2003–March 2007. Approximate location of equator is shown. (*Source:* Adapted from World Health Organization. © WHO 2007. All rights reserved.)

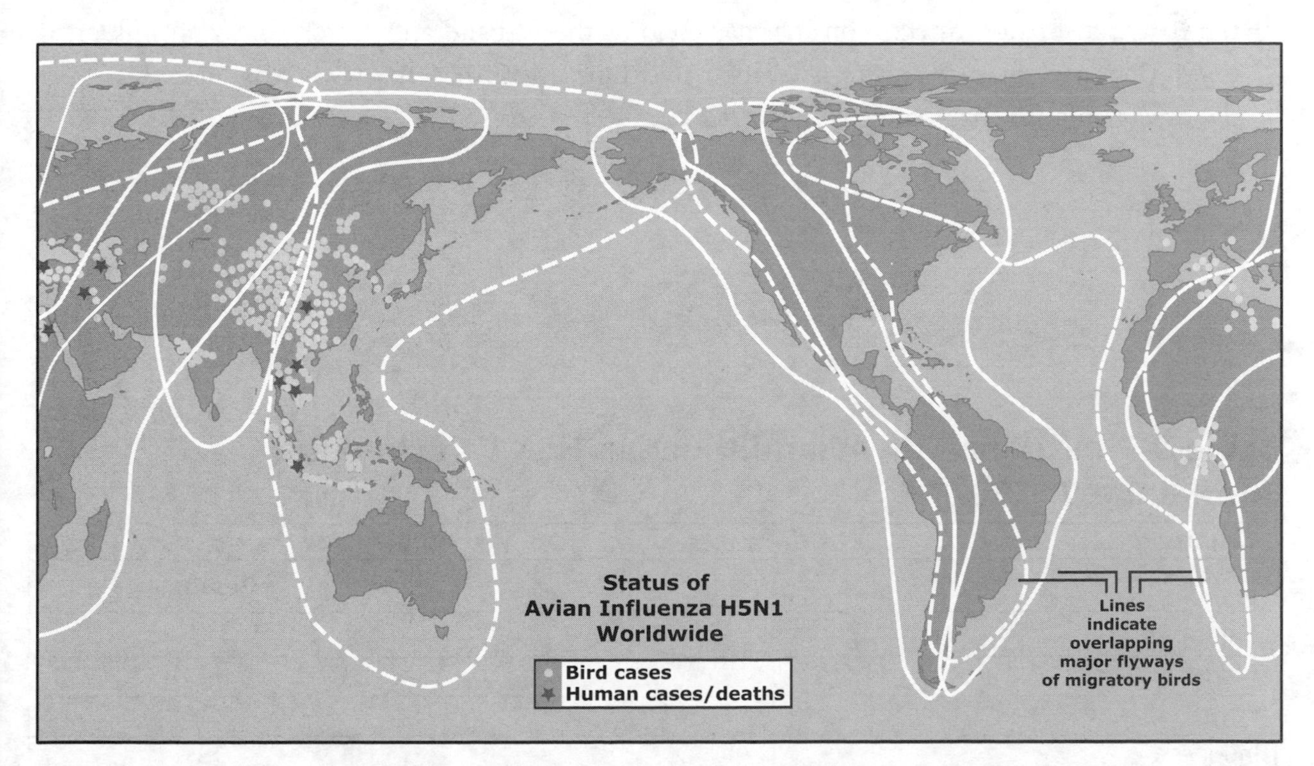

Figure 9.2c Global migration flyways. (*Source:* Adapted from Jan Conroy/UC Davis graphic. *Sources:* AI outbreaks: OIE, FAO and government sources; Flyways: Wetlands International.)

1. Examine Figure 9.2a. In 3–4 sentences, describe the general pattern of H5N1 avian flu that is shown in this map. Tell what continents and approximate latitudes are affected, where it seems to be most common in poultry, and where wild birds are the only cases reported so far.

2. Next examine Figure 9.2b. How does the density of human cases compare with the distribution of the H5N1 flu in poultry? In wild birds?

3. What inferences can you draw from these data about how avian flu is transmitted to humans? Explain each one, referring to the data you used in developing that inference.

4. Update Figure 9.2b with the most recent data from the World Health Organization. Place the new information directly on the map. (Go to the website and search for avian influenza cases. Data in Table 9.2 are from WHO, dated June 7, 2007. http://www.who.int/csr/disease/avian_influenza/country.)

Table 9.2 Incidence of Avian Influenza H5N1 in Humans, April–June 7, 2007

Country	2007	
	Cases	**Deaths**
Cambodia	1	1
China	1	1
Egypt	5	1
Indonesia	18	13
Total	25	16

5. Use the updated Figure 9.2b to calculate the death rate due to bird flu for Indonesia, Vietnam, China, Nigeria, and Egypt.

   Death rate = (number dead/total cases) × 100%

   Round to nearest 0.1%.

Indonesia ________________ Vietnam ________________ Nigeria ________________

China ____________________ Egypt ________________

6. If you were planning a trip to one of the countries listed in number 5, would you rather be told the death rate or the number of cases and deaths in each country? Explain.

7. Egypt reported its first cases of avian influenza in humans in December 2006. In the first 3 months of 2007, there were 14 more human cases of avian flu, with three deaths in that quarter year. Like people in much of the world, most Egyptian households keep small numbers of chickens. However, a tradition unique to Egypt is that its poultry are fed mouth to mouth by women who first chew grain and then blow the powdered feed into the mouths of their birds.

   What does this information further suggest about the ways avian influenza H5N1 spreads to humans? What additional information would you want in order to confirm your idea?

8. As you know, many wild birds fly to different parts of the earth as they follow their annual cycles of migration, mating, nesting, and overwintering. Examine the major flyways shown

in Figure 9.2c. In a few sentences, describe the patterns you see in these flyways. Include descriptors such as the degree to which they overlap, their general directional orientations, and their extent.

9. Given what you now know about the location of avian influenza and bird migration, make an inference about where you think this disease is most likely to be first detected in the United States. Explain your reasoning.

10. From which part of the world do you think avian flu came to Nigeria? Explain.

11. Extending this investigation, visit one of the many interactive maps on avian influenza. Two particularly good ones are offered by the British Broadcasting group in the United Kingdom, http://news.bbc.co.uk/1/shared/spl/hi/world/05/bird_flu_map/html/1.stm, and MSNBC, http://www.msnbc.msn.com/id/12375868/from/ET/.

    a. Tell which map you viewed and at least two new bits of information you learned.

b. Write two questions that you are able to answer with the information on the interactive map that you could not answer with the static maps provided in this investigation.

III. Critical Reading: The Body's Defenses Against the Flu

Before delving further into this investigative case, you should be familiar with the structure and reproduction of viruses, especially the influenza virus. If you have not already read Chapter 19: Viruses, you should do so now as background.

This Critical Reading is focused mainly on Chapter 43: The Immune System. Read the Overview; read Concepts 43.1, 43.2, and 43.3 through Active and Passive Immunization; and in 43.4, emphasize Acquired Immune System Evasion by Pathogens.

Innate Immunity: All Animals

1. Once a pathogen (a biological agent that causes disease) manages to get past an animal's physical barriers, there is a very good chance that it will be detected by the animal's immune system. At the molecular level, how does an animal detect the presence of a pathogen and determine that it is an intruder and nonself?

2. A single virus particle of influenza A is insufficient to initiate disease in humans. It is estimated that between 100 and 1,000 influenza A virus particles are necessary to cause the flu in an individual. A single droplet sneezed from an infected person is likely to contain sufficient virus particles to initiate the disease.

   Explain why a single virus particle is insufficient. To do this, describe both the barrier and the cellular mechanisms of innate immunity that could play a role in stopping viral particles from initiating and then spreading flu within a vertebrate host.

A recent summary of clinical findings in people infected with the avian H5N1 influenza A showed that levels of interferons, various interleukins, tumor necrosis factor (TNF), and monocyte attractant protein were higher in patients who died than in those who survived. These strong cellular responses of the innate immune system may actually have contributed to the multiorgan failure and sepsis seen in many patients (Writing Committee of the World Health Organization, 2005). By elevating the innate immune response to dangerous levels, it appears the H5N1 avian influenza turned the host's innate defenses against the host in some patients, resulting in several deaths.

Acquired Immunity: Vertebrates Only

In vertebrates, once the innate immune response fails to prevent host invasion, the acquired immune response takes over. In contrast to the more general innate immunity, acquired immunity is a response to specific pathogens. Unfortunately, it takes from several days to 2 weeks for the acquired response to be fully effective. Records from the U.S. Navy revealed that when the 1918 flu was diagnosed on any ship, the entire crew was quarantined onboard and virtually all crew members got sick. Many of the men died despite their own immune responses and medical treatment.

3. In the acquired immune response, two types of lymphocytes, B cells and T cells, are the key players.

Review Concepts 43.2 and 43.3 and Figures 43.9 and 43.10.

a. Compare and contrast the antigen receptors of B and T cells.

b. B and T cells can only recognize antigens when they are in specific locations. Where are the antigens located that B cells can detect?

c. Where are the antigens located that the two kinds of T cells can detect? Include the role of the MHC in your answer.

4. Examine Figure 43.16 and focus on the acquired immune system events following first exposure to an antigen. This was the case for the sailors, coal miners, and everyone else infected in 1918.

   a. How are the innate and the acquired immune systems linked? How does one system inform the other of an invasion by a foreign particle?

   b. In the acquired immune response, which cells undergo clonal selection once their receptors have joined with one of the antigens?

   c. Why is clonal selection a key event in whether the acquired immune response is successful?

   d. B plasma cells are effector B cells that fight off the infection differently than cytotoxic T cells do. Explain.

e. Although not a factor in disease caused by influenza A, what would be the effect on the acquired immune response if the helper T cells were removed from the system? (The virus causing the current AIDS pandemic targets such T cells.)

5. Immunization against common strains of human-to-human transmissible influenza A is needed every year because the seasonal flu mutates readily. In the case of the 1918 flu, immunization was poorly understood and not available. For a possible H5N1 pandemic, immunization will play a significant role.

   In the acquired immune system, how does an immunization protect a person from a disease like flu?

IV. Internet Activity: Influenza in the Media

The communication of biological information is an essential service for our global society. It is not unexpected that a current concern like the avian flu is frequently reported in the media, but the 1918 flu is also still in the news.

In early 2007, several articles cited Sir Mark Sykes, a victim of the Spanish flu epidemic, who died at the age of 39 in a hotel room in February 1919 while attending the Paris peace conference following World War I. Because the British diplomat was buried in a sealed lead coffin, researchers were hopeful that well-preserved body samples could be obtained.

1. After the flu virus is retrieved from the diplomat's remains, what is likely to be done with the sample?

2. Considering that it is 90 years since the outbreak of the 1918 flu, are you surprised by the current scientific interest? (When is scientific investigation done?)

To consider how current avian flu biology is being communicated globally, you can use a search engine such as Google Image Search to look for images used to communicate with the public in countries where human cases of avian flu have been reported.

3. Choose three images that you believe are helpful in presenting bird flu facts. Record your information in the following chart.

URL	Description	Country

4. What kinds of organizations produce the websites containing the media that you have listed?

5. Briefly describe the biological images used and how each relates to the avian flu—for example, a masked and gloved (to prevent viral infection) worker examining poultry (potential carriers of H5N1).

6. Which of your three media choices would be the most useful for convincing your classmates that avian flu is a real biological concern? Explain.

7. If you were asked to produce a brochure to advise a segment of the public in your country about the risks of avian flu, what group would you select as the target audience? Explain your choice.

8. What images would you include in your brochure? Why?

9. Describe at least three examples of biological information you would include for this particular group.

V. Using Data to Explore Pandemic Flu (Past and Possible)

A. Working with Tables and Graphs on Mortality Statistics

Consider the data in Table 9.3.

Table 9.3 Average Age at Death in the United States, 1911–1919

Year	Male	Female
1911	50.9	54.4
1912	51.5	55.9
1913	50.3	55
1914	52	56.8
1915	52.5	56.8
1916	49.6	54.3
1917	48.4	54
1918	36.6	42.2
1919	53.5	56

Adapted from *Life Expectancy in the USA, 1900–98* (Noymer, 2007).

1. In 1915, a man could expect to live for____________ years. In 1918, this dropped to __________ years.

2. Between 1917 and the onset of the influenza pandemic in 1918, both male and female life expectancy dropped __________ years.

3. Construct a graph with 1911 to 1919 on the *x*-axis and 0 to 60 years of age on the *y*-axis. Show separate male and female life expectancy lines by connecting the points plotted for each year versus expected age at death.

Provide a suitable title for your graph.

4. Which is more effective at quickly conveying the impact of the 1918 influenza pandemic, the table or your graph? Why?

Now look at Table 9.4.

Table 9.4 U.S. Deaths per 100,000 Attributed to Influenza and Pneumonia, 1917–1918

Age	1917	1918
<1	2,944.5	4,540.9
1–4	422.7	1,436.2
5–14	47.9	352.7
15–24	78	1,175.7
25–34	117.7	1,998
35–44	193.2	1,097.6
45–54	292.3	686.8

Adapted from *Age-specific death rates (per 100,000), Influenza & Pneumonia, USA* (Noymer, 2007).

5. Which age group had the highest number of influenza- and pneumonia-related deaths in 1917? In 1918?

6. Another way of looking at the data is to consider the specific increase in the number of influenza- and pneumonia-related deaths in one age group between 1917 and 1918. You can calculate this by dividing the number of deaths in 1918 by the number of deaths in 1917. For instance, if you look at the age group <1 year, divide the number of deaths in 1918 (4,540.9) by the number of deaths in 1917 (2,944.5). The increase in deaths during 1918 is approximately 1.5 times the 1917 deaths.

   How many times greater is the 1918 death total than the 1917 death total for 5- to 14- year-olds?

7. Which two age groups experienced the highest increase in the number of influenza- and pneumonia-related deaths between 1917 and 1918?
   a. How many times greater is the 1918 total than the 1917 total for each of these two groups?

   b. If you were the same age you are now, how much more likely would you have been to succumb from an influenza- or pneumonia-related death in 1918 than in 1917?

8. Scientists were puzzled why the 1918 flu resulted in increased deaths in specific age groups. What was so unexpected?

B. Working with the SIR Model to Investigate Avian Influenza H5N1

Models help investigators ask questions as well as predict possible outcomes. In this activity, you will use an Excel model (Weisstein, 2007) to simulate disease spread throughout a population. The model divides the host population into three categories:

- Susceptible individuals (*S*)
- Infected individuals (*I*)
- Individuals who have recovered from infection (*R*)

This SIR model produces graphs that track movement into, out of, and between the S, I, and R categories over time. Careful observation enables you to consider how the disease interacts with its host populations. (*Note:* Each disease being modeled is based on specific characteristics that the model user sets.)

Scenario 1

Consider the following simulation results (Figure 9.3) based on a hypothetical scenario in which four poultry workers from Tennessee infected with an H5N1 strain are moved to a health facility with a population of 180.

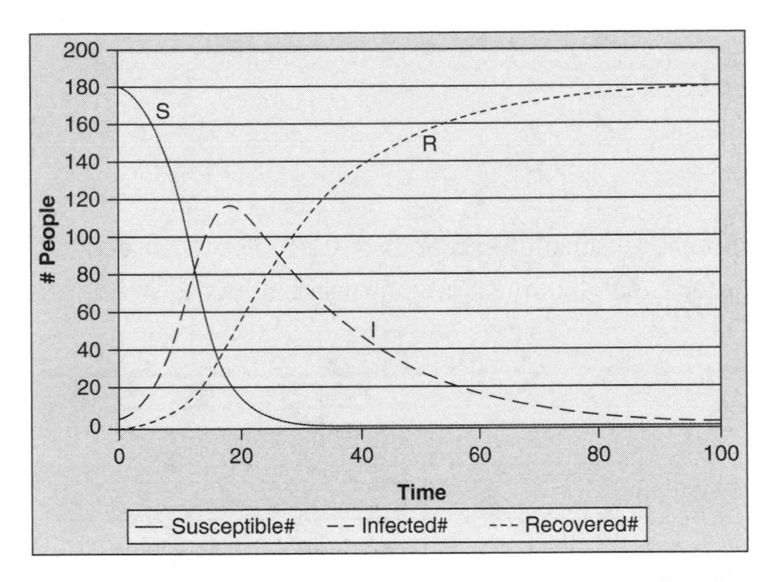

Figure 9.3 Simulation results for Scenario 1 of avian influenza.

1. Is this more likely to be a strain of the virus that is transmitted only from avian to human or a strain that is transmitted from human to human? Explain.

2. About how many individuals remained healthy? What is the transmission rate in this model? Transmission = R (recovered) / S (susceptible)

3. All infected individuals (I) eventually become recovered individuals (R). What does this tell us about the severity of the viral disease in this model—that is, what is the mortality rate observed here?

Scenario 2

This is the same hypothetical situation except that 40 people are isolated in a special ward at the facility before they can be exposed to the preceding avian influenza strain.

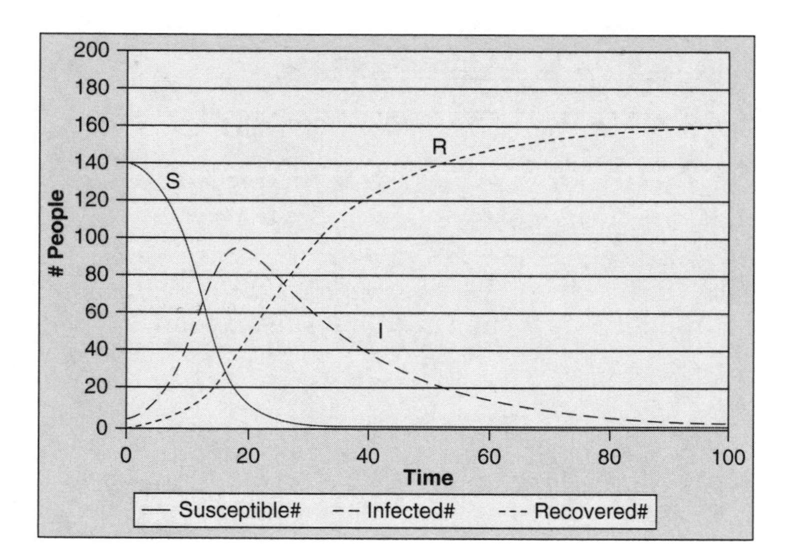

Figure 9.4 Simulation results for Scenario 2 of avian influenza.

4. The number of infected individuals (I) on Day 20 in Scenario 2 (see Figure 9.4) is __________ as compared to the number of infected individuals (I) on Day 20 in Scenario 1 (see Figure 9.3), which is ____________________.

5. The total number of recovered individuals (R) in Scenario 2 is __________ as compared to __________ in Scenario 1. What does this tell us about the efficacy of isolation in this model?

Scenario 3

In 1918, no one knew what caused influenza and vaccines were not available. Nevertheless, it was understood that coughing and sneezing contributed to the spread of the disease. At that time, Seattle public health officials required all passengers and employees of mass transit systems to wear masks (Figure 9.5). Spitting was also prohibited in many cities.

In the following simulations, consider the impact of individuals wearing masks during exposure to H1N1. These simulations involve a population of 200 hospital workers.

6. Predict generally what you'd expect to see in the SIR model results with respect to S, I, and R individuals. (Consider how these results would differ from not wearing masks.)

Figure 9.5 Street car conductor in Seattle not allowing passengers aboard without a mask. Record held at: The National Archives.

In the following simulations, let's reduce the transmission rate to approximately 10% of its previous value once the hospital initiates mask use. Exactly *when* the hospital starts using masks can dramatically affect the course of the epidemic, as the graphs in Figure 9.6 through Figure 9.8 show.

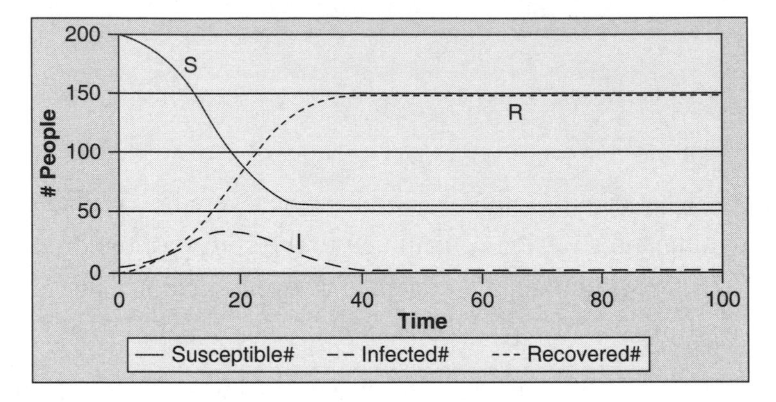

Figure 9.6 Simulation results for Scenario 3 of avian influenza, with masks used starting on Day 30, when the epidemic has already nearly run its course.

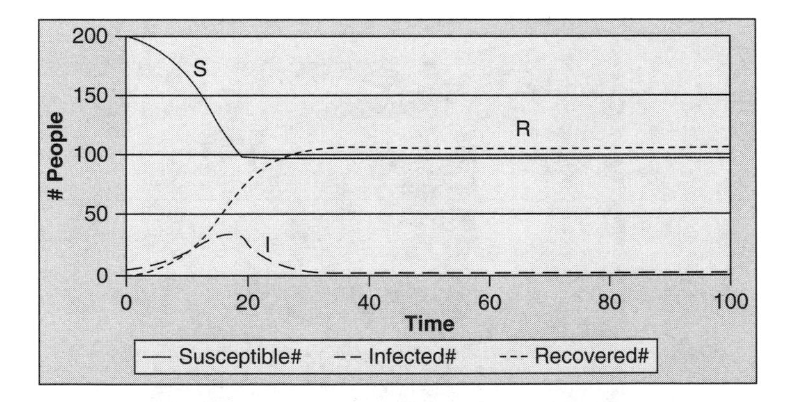

Figure 9.7 Simulation results for Scenario 3 of avian influenza, with masks used starting on Day 20, when the epidemic is at its peak.

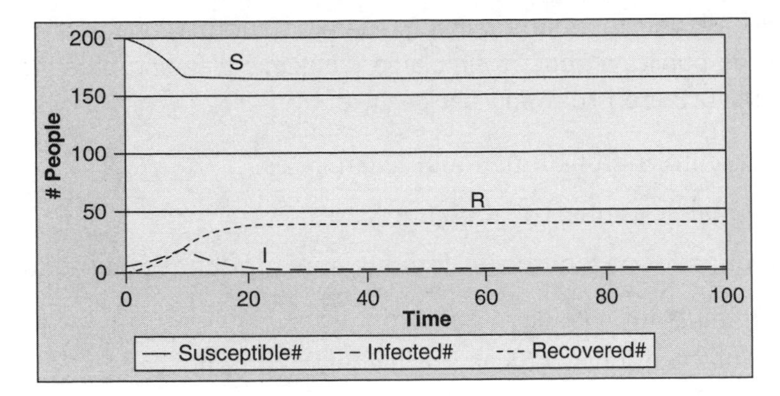

Figure 9.8 Simulation results for Scenario 3 of avian influenza, with masks used starting on Day 10, when the epidemic is still in its growth phase.

7. Using your observations of Figures 9.6 through 9.8, defend a policy that makes wearing masks mandatory during the early days of an epidemic such as the 1918 flu.

Additional Investigations

VI. Pandemic Planning

Although the H5N1 avian influenza A strain may not be the next virus that causes an epidemic around the globe, its severity and high mortality rate in humans have raised the level of alert.

Pandemics occur somewhat regularly, and planning is ongoing. International cooperative groups are already planning responses to a flu pandemic. At the same time, planning is proceeding at national, state, and local levels. Planning for a pandemic is complex and involves the participation of many stakeholder groups that have overlapping areas of concern and expertise. A list follows of some of the concerns and issues that need to be addressed in pandemic planning—in this case, for avian influenza.

Address one or more of these concerns from the perspective of one of the following stakeholder groups: public health (including CDC and WHO); medicine-pharmacology; media and public information; agriculture; local, state, and national governing bodies; department of state; consumers and taxpayers.

Your instructor may ask you to write a paper, prepare an informational poster or a 30-second radio spot informing the public, conduct a simulated pandemic planning meeting, or prepare some other form of response to these issues and perspectives.

1. Surveillance and monitoring of human and avian cases
2. Fostering international scientific collaboration
3. Developing international cooperation in limiting spread, defining acceptable enforcement
4. Prevention, limiting spread, and ethics
   a. Pharmacologic interventions—efficacy, availability, compliance
      i. Vaccination
      ii. Antivirals as prophylactics
   b. Nonpharmacologic interventions (NPIs)—efficacy and gaining compliance
      i. Community mitigation (school closures, restrictions on movements)
      ii. Quarantine and isolation of healthy and sick individuals
      iii. Travel restrictions locally and globally
      iv. Social distancing
      v. Use of masks and increased hygiene
5. Treatment and ethics
   a. Antiviral availability
   b. Hospital space limitations for isolation and treatment
6. Disposal of bodies if pandemic is extreme
7. Continuation of governance during a pandemic
8. Delivery of services when many are sick—priorities and ethics
   a. Water, food, energy
   b. Workforce reductions due to illness
9. Economic impact (short term) and ethics
   a. Confined animal feeding operations (CAFO)—placement and regulation
   b. Importers and exporters of other goods and services
   c. Pharmaceutical industry
   d. Health-insurance industry
   e. Health-care industry
10. Long-term effects on affected countries and ethics
    a. Economic effects
    b. Loss of personnel to pandemic

Some Suggested Resources Related to Pandemic Planning

Pandemicflu.gov

Avianflu.gov

National Governor's Association document on Preparing for Pandemic Influenza http://www.nga.org/Files/pdf/0607PANDEMICPRIMER.PDF (Note: This is a large file.)

Centers for Disease Control and Prevention (http://www.cdc.gov/flu/avian)

World Health Organization (http://www.who.int/en/)

VII. Open-Ended Investigations

The Excel-based SIR model used in Investigation V is freely available online at http://bioquest.org/icbl/casebook/avian. Consider developing your own scenarios (more virulent flu, different mitigation efforts, etc.) and seeing their effectiveness on the spread of disease.

References

deJong, Menno D., Tran Tan Thanh, Truong Huu Khanh, Vo Minh Hien, Gavin J. D. Smith, Nguyen Vinh Chau, Bach Van Cam, Phan Tu Qui, Do Quang Ha, Yi Guan, J. S. Malik Peiris, Tran Tinh Hien, and Jeremy Farrar. Oseltamivir resistance during treatment of influenza A (H5N1) infection. *NEJM,* 353(25):2667–672, 2005.

Hawkins, Nyoka. Teamus Bartley interview, Appalachia Oral History Project, Louie B. Nunn Center for Oral History, University of Kentucky Libraries, Accession No. 87OH191 App 114. Date June 14, 1987.

Kuiken, T., E. C. Holmes, J. McCauley, G. F. Rimmelzwaan, C. S. Williams, and B. T. Grenfell. Host species barriers to influenza virus infections. *Science,* 312(5772):394–397, 2006.

Noymer, A. Raw data set: *Age-specific death rates (per 100,000). Influenza and Pneumonia, USA.* http://www.demog.berkeley.edu/~andrew/1918/figure1.html (accessed April 2007). *Life expectancy in the USA, 1900–98.* http://www.demog.berkeley.edu/~andrew/1918/figure2.html (accessed April 2007).

Taubenberger, J. K., and D. M. Morens. 1918 influenza: The mother of all pandemics. *Emerging Infectious Diseases*, on return 12(1), January 2006. Online at http://www.cdc.gov/ncidod/EID/vol12no01/05-0979.htm.

Weisstein, A. E. SIR Modeling. In Biological ESTEEM: Excel Simulations and Tools for Exploratory, Experiential Mathematics. 2007 http://bioquest.org/esteem.

World Health Organization. Antigenic and genetic characteristics of H5N1 viruses and candidate H5N1 vaccine viruses developed for potential use as pre-pandemic vaccines. http://www.who.int/csr/disease/avian_influenza/guidelines/h5n1virus/en/index.html (accessed March 2007).

The Writing Committee of the World Health Organization (WHO) Consultation on Human Influenza A/H5. Avian Influenza A (H5N1) infection in humans. *NEJM,* 353(13):1374–385, 2005.

Maps

Figure 9.2a http://gamapserver.who.int/mapLibrary/Files/Maps/Global_SubNat_H5N1inAnimalConfirmedCUMULATIVE_20070329.png

Figure 9.2b http://gamapserver.who.int/mapLibrary/Files/Maps/Global_H5N1inHumanCUMULATIVE_FIMS_20070329.png

Figure 9.2c http://calag.ucop.edu/0603JAS/images/avianMap.jpg

CHAPTER 10

Shh: Silencing the Hedgehog Pathway

Shh: Silencing the Hedgehog Pathway

"I'm so relieved," Ann said as she plopped down in the coffee shop booth where her friend Delores was reading e-mail on her laptop.

"Oh, Ann!? What did the doctor say?" Delores asked.

"Well, I do have skin cancer, but it's not melanoma. It's basal cell something. Anyway, it's very common and easy to treat," Ann reassured her.

"Is it genetic?" Delores asked, "or does it have something to do with that nice tan you showed off during your teens and twenties?"

"Well no one else in the family has had skin cancer." Grimacing, Ann added, "It's more likely I'm paying for my tan."

After Ann left, Delores searched for "basal cell cancer" on the Web. She wondered how her friend ended up with skin cancer. She found a 2004 paper by Athar and colleagues that explored BCC (basal cell carcinoma) and the effect of UV radiation. BCC, the most common kind of cancer, was linked to problems with the hedgehog signaling pathway. Exposure to UV radiation was one way to impact the pathway.

"More questions than answers," Delores sighed. She looked up "hedgehog signaling pathway" in Wikipedia. She found that this pathway controls cell division and is important in early development. The pathway was first discovered in fruit flies with a mutation that made them shorter and especially bristly. The researcher thought the fly larvae looked like hedgehogs.

Delores returned to the Athar article. The researchers divided mice into two groups and then exposed them to UV radiation. One group was given a drug called cyclopamine, a known antagonist to the hedgehog pathway, in their drinking water, and the other group got plain water. The mice that got the cyclopamine had many fewer BCCs at the end of the experiment.

"I wonder if they will give Ann cyclopamine for her BCC?" Delores thought as she closed her laptop.

Figure 10.1 Using a laptop at a coffee shop.

CASE ANALYSIS

1. **Recognize potential issues and major topics in the case.** What is this case about? Underline terms or phrases that seem to be important to understanding this case. Then list **3–4** biology-related topics or issues in the case.

2. **What specific questions do you have about these topics?** By yourself, or better yet, in a group, list what you already know that is related to the case in the "What Do I Know?" column. List questions you would like to learn more about in the "What Do I Need to Know?" column.

What Do I Know?	What Do I Need to Know?

3. Put a check mark by **1–3** questions or issues in the "What Do I Need to Know?" list that you think are most important to explore.

4. **What kinds of references or resources would help you answer or explore these questions?** Identify two different resources and explain what information each resource is likely to give that will help you answer the question(s). Choose specific resources.

Core Investigations

I. Critical Reading: Cell Signaling Pathways

You should be familiar with the structure and function of proteins that have active sites, such as enzymes or antibodies. To complete this investigation, you should read Chapter 11: Cell Communication (specifically, Concepts 11.1 and 11.4) and Chapter 18: Regulation of Gene Expression (specifically, Concepts 18.4 and 18.5).

1. What is cancer? (Hint: Use of multiple sources for this definition, such as Cancerquest [http://www.cancerquest.org] in addition to the text, is recommended.)

2. What are some of the causes of cancer?

3. Interpret the graph in Figure 10.2 by answering the following questions.

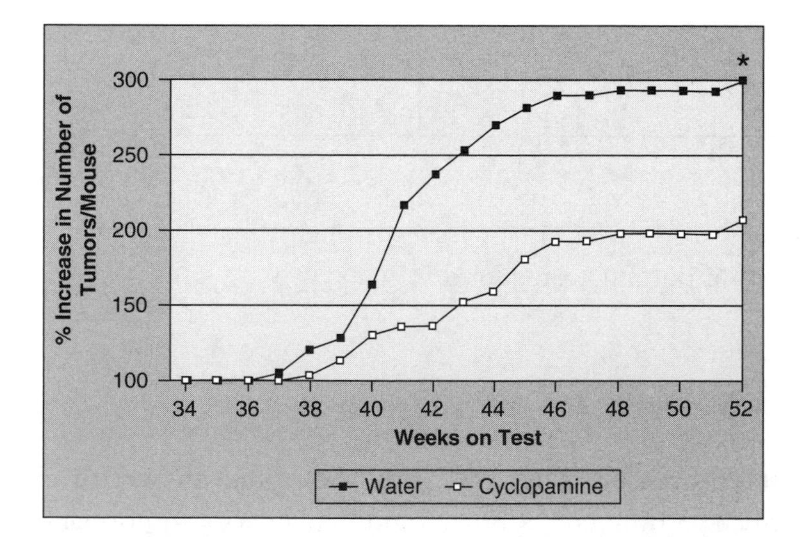

Figure 10.2 Effect of cyclopamine on BCC tumor formation in UVB-irradiated mice. (After Athar et al., 2004) (*Note:* The asterisk means the differences between the two treatments are statistically significant.)

a. On the basis of the shape of the curves, explain the patterns of tumor production in control and experimental mice in weeks 34–52.

b. What is the overall percentage increase in tumors for control versus experimental mice?

4. Use Figure 10.3 to answer the next two questions.

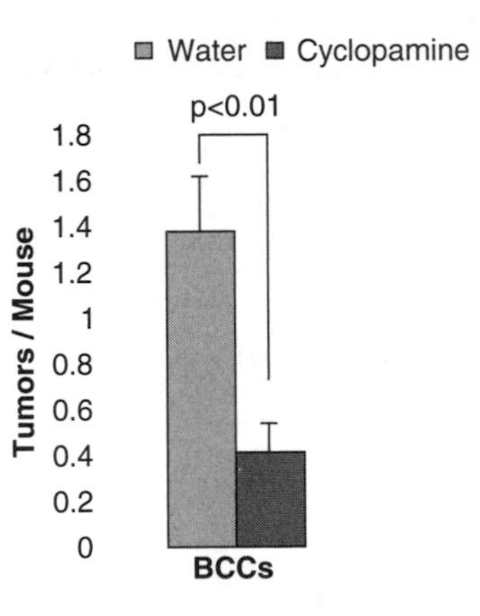

Figure 10.3 Average number of tumors per irradiated mouse with and without cyclopamine. (After Athar et al., 2004)

a. How effective was cyclopamine in treating BCC in the mice?

b. Which of these two data formats—the bar graph or the line graph—would be more effective in presenting the results of the experiment to the public? Which would be more effective for other scientists?

The hedgehog signaling pathway plays a crucial role in the development of many animal embryos. In addition, the hedgehog pathway controls regeneration of short-lived adult tissues, such as those in skin and blood. When the hedgehog pathway is active, transcription of proteins occurs in target cells followed by rapid division of those cells. The hedgehog pathway is also active in BCC and several other cancers.

The *hedgehog* gene family codes for signaling proteins that serve as ligands binding to receptors in nearby target cells. These proteins activate the hedgehog pathway in the target cells. The hedgehog pathway in the target cells has two membrane proteins named Patched (Ptch) and Smoothened (Smo), as well as several intracellular proteins.

When Shh (Sonic hedgehog) ligand binds to Ptch, then Smo is activated, the signal is transduced, and transcription and cell division result. In the absence of the hedgehog signaling protein, Ptch inhibits Smo, no signal is sent to the intracellular components of the hedgehog pathway, and thus transcription and cell division do not occur. Smo and the subsequent intracellular pathway may also be turned on by mutations that inactivate Ptch.

5. Is the hedgehog signaling pathway a local or long-distance type of signaling? Explain.

6. Examine Figure 10.4 and identify which molecules are involved in reception, transduction, and response in the hedgehog pathway.

7. The mechanism of the activation of Smo by the hedgehog ligand binding to Ptch is not completely understood. However, the model shown in Figure 11.11 in the text shows a pathway with two membrane proteins, similar to the arrangement of membrane proteins in the hedgehog pathway. In this model, cell signaling is involved when the ligand binds to the first receptor protein, activating the G protein. The G protein then activates the second membrane protein, which transduces the signal to the interior of the cell. Explain how this mechanism might be applied to the two membrane proteins in the hedgehog signaling pathway.

8. As scientists evaluate new data, they frequently have to revise their models. Because we know that Ptch is an inhibitor of Smo and G protein is not involved, revise the model in Figure 11.11 to incorporate this new information.

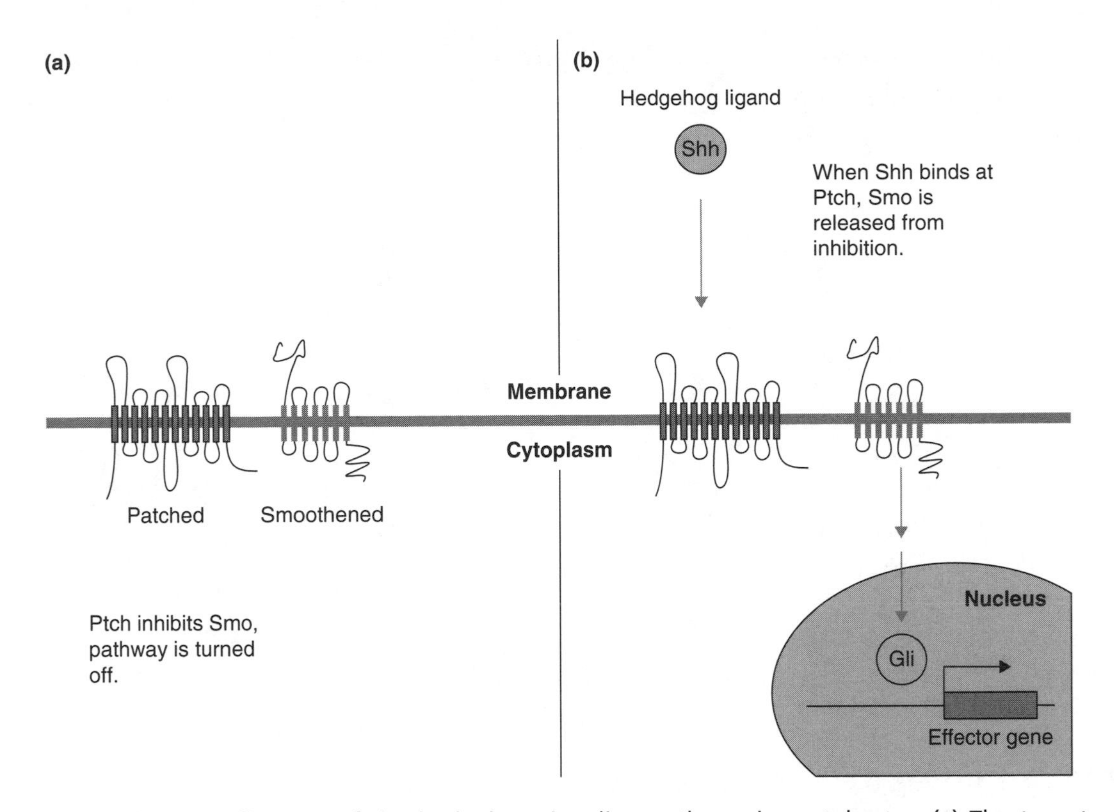

Figure 10.4 Schematic diagram of the hedgehog signaling pathway in vertebrates. (a) The target cell without the hedgehog ligand. Patched and Smoothened are transmembrane proteins embedded in the plasma membrane. Patched inhibits Smoothened and the pathway is turned off. (b) When the hedgehog ligand Shh joins with Patched, Smoothened is released from inhibition and the pathway is turned on. (Weitzman, 2002)

9. Cyclopamine is a known antagonist of Smo. Describe how cyclopamine reduces the number of BCCs in UVB-irradiated mice.

The hedgehog signaling pathway is active in the early embryo during development of the neural tube, motor neuron specification, left-right symmetry, body plan, limbs, and retinas (Matlack et al., 2006).

In the 1950s, sheep feeding on the corn lily (Veratrum spp.) in mountain pastures gave birth to a number of lambs with only one eye. The number of cyclopean lambs (named for the one-eyed Cyclops of Greek mythology) were explained when the compound later named cyclopamine was discovered in the corn lily. To see an image of cyclopia in sheep, go to http://teratology.org/jfs/NaturalTeratogens.html.

10. Explain how a failure to have cell division occur at a critical time during development could lead to lambs with one eye. See Concept 18.4 on critical events in the development of left-right symmetry and body plan.

II. Phylogenetics of the Hedgehog Gene Family

Nobel Prize researchers Christiane Nüsslein-Volhard and Eric Wieschaus investigated fruit fly mutations in order to make sense of the role of genes active in the development of fly embryos. They mutated one gene—later named *hedgehog*—that resulted in dense spines in shortened fly larvae.

Homologous *hedgehog* genes were later discovered in vertebrates. After these genes were sequenced in several different kinds of animals, they were compared and used to determine phylogenetic relatedness. If you have not yet studied phylogenetic classification, you may want to read Chapter 26 before completing this investigation. Consider the phylogram in Figure 26.12 in your text as you answer the following questions.

1. What species is used as the outgroup for the *hedgehog* gene in this phylogram? Provide a reason for using this species.

2. What does the phylogram tell us about the *hedgehog* gene in mammals and birds as compared to the *hedgehog* gene in mammals and amphibians?

Consider the ultrametric tree in Figure 26.13 in your text as you answer the following questions.

3. Does this ultrametric tree provide information about the rate of change in the *hedgehog* gene for these animal groups? If not, what information can be inferred from this tree?

4. List all the pairs of animal groups that share a more recent common ancestor than humans and birds.

The field of developmental biology is changing as scientists use molecular and biological approaches to investigate evolution questions. Since the discovery of classes of conserved regulatory genes or *Hox* genes, the new phrase "evo devo" has been used to refer to the science of evolutionary developmental biology. (See Concept 21.6 in your text.)

5. How could phylogenetic and ultrametric trees help inform researchers interested in designing experiments to study the hedgehog pathway?

After viewing the *hedgehog* phylogram in Figure 26.12, two researchers decide to look more closely at the relationship between hedgehog proteins found in animals. They obtain sequence information from two invertebrates and two vertebrates and choose to limit their study to a portion of the hedgehog protein produced by a highly conserved region of the *hedgehog* gene. The hedgehog proteins were produced by the gene Hh (hedgehog) in invertebrates and Shh (sonic hedgehog) in vertebrates.

The following amino acid sequences were produced from a conserved region of a gene in the *hedgehog* family of genes.

Scorpion Hh	DGPHAINSLH	YEGRAVDITT	SDRDRSKYGM	LARLAVDAGF	DWVYYESRAH	IHCSVKSESA
Human Shh	DGHHSEESLH	YEGRAVDITT	SDRDRSKYGM	LARLAVEAGF	DWVYYESKAH	IHCSVKAENS
Octopus Hh	QGHHAPTSLH	YEGRAVDITT	SDRVRSRYGM	LARLAVEAGF	DWVYYESRSH	IHCSVRSDSL
Chicken Shh	DGHHSEESLH	YEGRAVDITT	SDRDRSKYGM	LARLAVEAGF	DWVYYESKAH	IHCSVKAENS

6. Which of these organisms are invertebrates and to what phylum does each belong?

7. Which are vertebrates and to what class does each belong?

8. Based on the preceding limited amino acid sequences, which animal has the hedgehog protein most like the one in the human?

9. Which organism has the most differences in the amino acid sequence compared to the human hedgehog protein sequence?

10. Would you have predicted the answers to questions 8 and 9? Why or why not?

III. Critical Reading: Stem Cells and Gene Expression

Chapters 18 and 20 explore the basic question of how cells with the same DNA can become different cell types. Cell biologists cite differential gene expression in cells as the explanation. They are working with stem cells to better understand the regulation of gene expression in both developing and adult organisms. Cell signaling pathways play a critical role in this gene regulation, but it is important to note that pathways such as the hedgehog pathway have different roles in embryonic and adult stem cells.

With the exception of gametes, a complete set of chromosomes is found in all cells in the human body. However, not all genes are expressed in each cell. The proteins necessary for cell function depend on the location and function of a particular cell in the body as well as the specific conditions the cell confronts during its survival in the body.

1. Specify the location of a cell in your body that:
   a. contains genetic information on eye color and the production of insulin.

   b. expresses eye color.

   c. produces insulin.

Over the last century, cell determination in the developing embryo has been closely observed. New methods and tools enable modern scientists to probe this process at the molecular level.

2. What is the molecular definition of *determination*?

3. What molecules provide the earliest evidence that a cell is committed to a particular cell fate?

Stem cells are distinct from most cells in animals because they retain the ability to divide and remain relatively undifferentiated. Under certain conditions, however, stem cells divide and a subset of the new cells differentiates into specific cell types.

4. What are the major differences between stem cells found in embryonic tissue and those found in adult tissues?

The current interest in stem cells for regulation of gene expression is tied to providing potential new therapies for treatment of diseases such as the use of hedgehog pathway antagonists.

Current research with adult stem cells has provided some unexpected results (Figure 10.5).

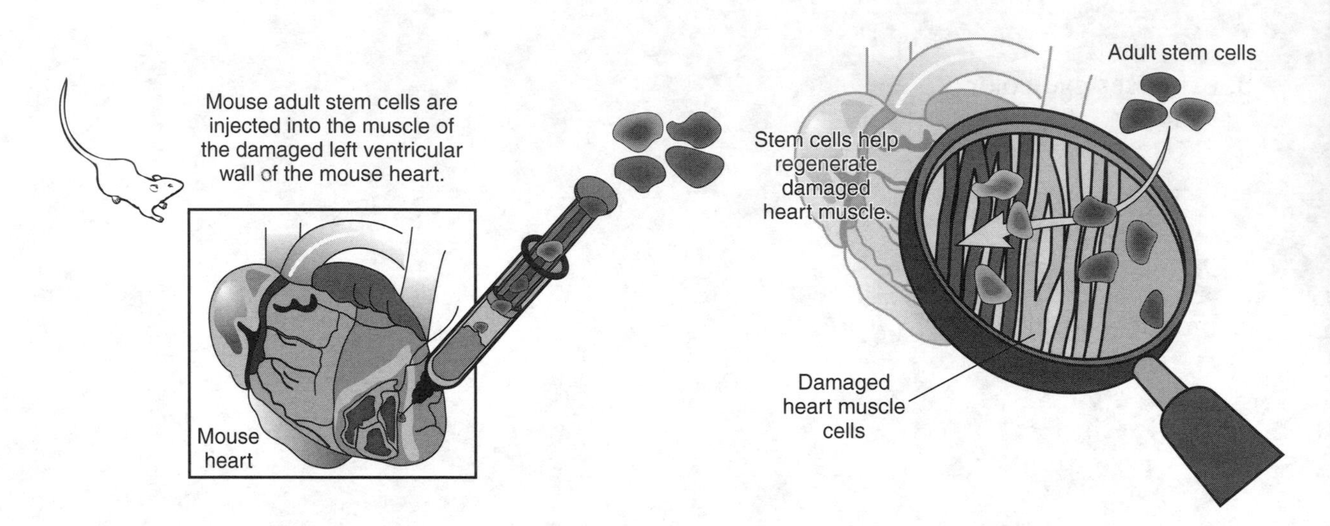

Figure 10.5 Mouse bone marrow stem cells injected into a damaged mouse heart resulted in new heart tissue. The bone marrow stem cells appear to secrete factors that promote regeneration.

Reference for image:
Stem Cell Basics: What are the potential uses of human stem cells and the obstacles that must be overcome before these potential uses will be realized? 2006. In *Stem Cell Information* (World Wide Web site). Bethesda, MD: National Institutes of Health, U.S. Department of Health and Human Services (cited Tuesday, June 19, 2007). Available at http://stemcells.nih.gov/info/basics/basics6.

5. What kinds of cells can adult bone marrow cells give rise to?

6. Describe why human adult stem cells from bone marrow are used to treat patients who have undergone radiation treatments.

7. Provide an example of how treatment with cultured embryonic stem cells could be used to supply cells for the repair of damaged or diseased organs in human patients.

8. Use of adult stem cells is well accepted; however, these cells have limited use as donor cells. Human embryonic stem cells have greater potential uses in a wider variety of tissues; however, the use of embryonic stem cells raises ethical and political issues. Identify two concerns an individual might have with the use of embryonic stem cells.

IV. Investigating the Hedgehog Pathway: Antibodies as Research Tools

Scientists ingeniously design research tools based on *in vivo* processes of biological systems. For example, PCR is a technique utilizing the enzyme DNA polymerase to initiate the synthesis of a minuscule DNA sample. Likewise, Western blots and immunohistochemistry are techniques utilizing the highly specific binding of antibodies with target molecules to act as molecular probes in cells and tissues.

1. Explain how an antibody is able to recognize a specific antigen. (Include an explanation of an epitope in your answer.)

Antibody Techniques

Antibodies can be used to find, bind, and tag a specific molecule of interest. Antibodies are Y-shaped molecules, with the tips of the Y containing unique amino acid sequences that bind the antigen. These variable portions of antibody molecules convey the high specificity for a target molecule. The large tail or base of the Y is much less variable. In fact, all antibodies within a species have tail regions that are very similar in sequence and shape. By inserting compounds that fluoresce, produce radiation, or produce a color change in the tail region of these molecules, researchers can use antibodies as marker molecules.

Often, researchers use two different antibodies: a primary antibody for targeting the molecule of interest and a secondary antibody with active sites to bind the primary antibody's tail and act as a marker.

Western blots are used to detect the presence of a known protein in a given sample. The proteins are first separated by molecular weight using gel electrophoresis. Next, the proteins are transferred from the gel to a nitrocellose membrane in a process called blotting. The nitrocellulose membrane is then incubated with a primary antibody that combines with the protein of interest. Then an enzyme coupled with a secondary antibody is used to produce a detectable color change when the protein of interest is present. The intensity of the color change indicates the quantity of the protein. *Immunohistochemistry* uses antibodies to detect the presence of specific molecules, usually proteins, within tissues and cells. Thin sections of a biological sample are fixed to a glass slide, incubated with primary and secondary antibodies, and examined microscopically. Secondary antibodies used in immunohistochemistry are frequently fluorescent, in which case a fluorescent microscope is used to read the results.

2. Briefly describe what you can learn about a target protein by using each of these two techniques.

In 1996, it was discovered that a mutation in the *Patched* gene in the hedgehog pathway was involved in almost all cases of basal cell nevus syndrome, a rare hereditary syndrome of birth defects and multiple BCC starting early in life. This autosomal recessive mutation was identified in families affected by the syndrome. The *Patched* mutation resulted in a nonfunctional Patched protein.

3. Refer to the hedgehog pathway diagram (Figure 10.4) and explain what happens when the Patched protein is nonfunctional.

4. How could this lead to cancer?

As soon as the hedgehog pathway was implicated, researchers began looking at inhibitors that might serve as chemotherapy for this common cancer. Cyclopamine, the plant teratogen known to interfere with the hedgehog signaling pathway in early development, showed potential as a cancer treatment.

The exact mechanism by which cyclopamine inhibits hedgehog pathway signaling has been a topic of controversy. One hypothesis was that cyclopamine prevented the secretion of Shh from Shh-producing cells. The protein was expressed but without its normal cholesterol addition. As a result, Shh could not leave the cell to act as a signaling protein.

5. If you were asked to test this hypothesis, which technique do you think would be more useful—Western blots or immunohistochemistry? Explain your choice.

The Following Experiment was Designed to Test the Effect of Cyclopamine on Shh Secretion

Chick embryos in embryonic Day 3 (cells known to secrete Shh) are divided into cyclopamine-treated groups and control groups. After an established exposure time, embryos from both groups are sacrificed and thin sectioned for immunohistochemistry testing. Samples are incubated with primary, then secondary, antibodies per an established protocol. Under fluorescent microscopy, tissues are assessed for the presence and location of Shh.

Table 10.1 Antibodies for Hedgehog Proteins

Order Number	Protein	Tissue Specificity	Antibody Type and Source
1223	Hh	Drosophila	*Anti-Hh* goat, polyclonal
1224	Hh	Drosophila	*Anti-Hh* rabbit, polyclonal
2011	Ptc	Drosophila	*Anti-Ptc* goat, polyclonal
2624	Ptch	Mouse, rat, human, chicken	*Anti-Ptch* rabbit, polyclonal
2680	Ptch	Mouse, rat, human	*Anti-Ptch* goat, polyclonal
2681	Ptch	Mouse, rat, human	*Anti-Ptch* goat, polyclonal
2626	Ptch	Mouse, rat, human, chicken	*Anti-Ptch* rabbit, polyclonal
4235	Shh	Mouse, human	*Anti-Shh* mouse, monoclonal
4257	Shh	Human	*Anti-Shh* rabbit, polyclonal
4278	Shh	Mouse, human	*Anti-Shh* rat, monoclonal
4279	Shh	Mouse	*Anti-Shh* goat, polyclonal
4284	Shh	Mouse, rat, human, primate, chicken, cat	*Anti-Shh* goat, polyclonal
4286	Shh	Mouse, rat, human, zebrafish, Xenopus	*Anti-Shh* rabbit, polyclonal
3511	Smo	Drosophila	*Anti-Smo* mouse, monoclonal
6766	Smo	Mouse, rat, human	*Anti-Smo* rabbit, polyclonal
6788	Smo	Mouse, rat, human	*Anti-Smo* goat, polyclonal
6789	Smo	Drosophila	*Anti-Smo* goat, polyclonal

Based on a page from a research supply company catalog.

6. Consider the antibodies for hedgehog proteins listed in Table 10.1. Which one of these primary antibodies would you choose for the experiment described earlier? Why?

7. The secondary antibody includes a fluorescent marker in its tail region. Why?

The results of this experiment showed no difference between the test and the control specimens in the amount of fluorescence inside and outside the cells.

8. What can you conclude from these results?

New evidence suggests that cyclopamine's effect on the pathway may be caused by inhibiting Smoothened. Therefore, even in the presence of Shh, no message is transduced to the nucleus by Gli, and therefore there is no cellular response.

9. Consider the use of cyclopamine as a chemotherapeutic agent in cases of spontaneous BCC. Researchers discovered that this cancer results from a mutation of *Patched*. If cyclopamine was approved for human use, would you recommend it for these cases of BCC? Why or why not?

Additional Investigations

V. Open-Ended Investigations

To learn more about the hedgehog pathway, you could explore the Hedgehog Pathway Signaling Database (Ramirez-Weber, 2006), which contains relevant information, images, and references to research articles. You may wish to form a group to develop a proposal for a new investigation. For example:

- Explore a known hedgehog pathway antagonist other than cyclopamine.
  - How does this antagonist disrupt the pathway?
  - Does the antagonist have potential as a chemotherapeutic drug?
  - What organism produces this antagonist molecule and how is it useful in that organism?
- Choose one of the molecules in the pathway and compare the genes using Biology Workbench.

Note: Your instructor could set up a proposal peer review process in your class, simulating what is done by major funders such as the National Science Foundation and the National Institutes of Health.

References

Anonymous. Stem cell basics: What are the potential uses of human stem cells and the obstacles that must be overcome before these potential uses will be realized? In *Stem Cell Information* [World Wide Web site]. Bethesda, MD: National Institutes of Health, U.S. Department of Health and Human Services, 2006. http://stemcells.nih.gov/info/basics/basics6 (accessed June 19, 2007).

Athar, Mohammad, Chengxin Li , Xiuwei Tang, Sumin Chi, Xiaoli Zhang, Arianna L. Kim, Stephen K. Tyring, Levy Kopelovich, Jennifer Hebert, Ervin H. Epstein Jr., David R. Bickers, and Jingwu Xie. Inhibition of smoothened signaling prevents ultraviolet B-induced basal cell carcinomas through regulation of Fas expression and apoptosis. *Cancer Research* 64:7545–7552, 2004. http://cancerres.aacrjournals.org/cgi/content/full/64/20/7545#F2 (accessed September 2007).

Cancerquest. http://www.cancerquest.org (accessed September 2007).

Chen, James K., Jussi Taipale, Michael K. Cooper, and Philip A. Beachy. Inhibition of hedgehog signaling by direct binding of cyclopamine to Smoothened. *Genes and Development,* 16:2743–2748, 2002. http://www.genesdev.org/cgi/content/full/16/21/2743 (accessed September 2007).

Johnson, R. L., A. L. Rothman, J. Xie, L. V. Goodrich, J. W. Bare, J. M. Bonifas, A. G. Quinn, R. M. Myers, D. R. Cox, E. H. Epstein Jr., and M. P. Scott. Human homolog of Patched, a candidate gene for the basal cell nevus syndrome. *Science,* 272:1668–1671, 1996.

Kumar, S., K. Balczarek, and Z. Lai. Evolution of the hedgehog gene family. *Genetics,* 142:965–972, 1996.

Matlack, David. Private communication regarding the use of immunohistochemistry and antibodies, June 2007.

Matlack, D., P. Pape-Lindstrom, and S. Webb. Hedgehog-emony: BioQUEST Complex Data Sets Workshop Project. http://bioquest.org/summer2006/workshop_forms/project_template.php?project_id=251 (accessed June 21, 2007).

Ramirez-Weber, F. A. The Hedgehog Signaling Pathway Database. NIH RIMI 5P20-MD000262. San Francisco State University. http://hedgehog.sfsu.edu/ (accessed July 3, 2007).

Weitzman, J. G. Agonizing hedgehog. *Journal of Biology,* 1:7, 2002. Online at http://jbiol.com/content/1/2/7 (accessed September 2007).